174 Topics in Current Chemistry

Molecular Similarity II

Editor: K. Sen

With contributions by
P. J. Artymiuk, A. Kumar, P. C. Mishra,
I. Pálinkó, A. R. Poirette, R. Ponec,
D. W. Rice, Y. Takahashi, G. Tasi,
P. Willett

With 55 Figures and 12 Tables

Springer-Verlag Berlin Heidelberg GmbH

This series presents critical reviews of the present position and future trends in modern chemical research. It is addressed to all research and industrial chemists who wish to keep abreast of advances in their subject.

As a rule, contributions are specially commissioned. The editors and publishers will, however, always be pleased to receive suggestions and supplementary information. Papers are accepted for "Topics in Current Chemistry" in English.

Library of Congress Catalog Card Number 74-644622

ISBN 978-3-662-14889-1   ISBN 978-3-540-49040-1 (eBook)
DOI 10.1007/978-3-540-49040-1
© Springer-Verlag Berlin Heidelberg 1995

Originally published by Springer-Verlag Berlin Heidelberg New York in 1995.
Softcover reprint of the hardcover 1st edition 1995

Typesetting: Macmillan India Ltd., Bangalore-25
SPIN: 10474227   51/3020 - 5 4 3 2 1 0 - Printed on acid-free paper

Guest Editor

Prof. Dr. *Kali Das Sen*
University of Hyderabad
School of Chemistry
Central University P.O.
Hyderabad 500 134, India

Editorial Board

Attention
all "Topics in Current Chemistry" readers:

A file with the complete volume indexes Vols.22 (1972) through 173 (1995) in delimited ASCII format is available for downloading at no charge from the Springer EARN mailbox. Delimited ASCII format can be imported into most databanks.

The file has been compressed using the popular shareware program "PKZIP" (Trademark of PKware Inc., PKZIP is available from most BBS and shareware distributors).

This file is distributed without any expressed or implied warranty.

To receive this file send an e-mail message to:
SVSERV@VAX.NTP. SPRINGER.DE
The message must be:"GET/CHEMISTRY/TCC_CONT.ZIP".

SVSERV is an automatic data distribution system. It responds to your message. The following commands are available:

HELP	returns a detailed instruction set for the use of SVSERV
DIR (name)	returns a list of files available in the directory "name",
INDEX (name)	same as "DIR",
CD <name>	changes to directory "name",
SEND <filename>	invokes a message with the file "filename",
GET <filename>	same as "SEND".

For more information send a message to:
INTERNET:STUMPE@SPINT. COMPUSERVE.COM

Preface

In the past, similarity, or the lack of it, has been studied qualitatively in almost every discipline of knowledge. Recently, chemists have devoted considerable attention to devising several useful quantitative measures which have made the similarity concept of fundamental importance in molecular engineering. The present two volume monograph on 'molecular similarity' is an attempt to cover some of the main conceptual and computational development in this rapidly growing multidisciplinary area of research. It is hoped that several other quantum chemists (density-and wave-functional) will assist in refining the molecular similarity measures so as to make them uniformly applicable to small and large molecules and, equally importantly, contribute to defining other new measures.

I take this opportunity to place on record my deep sense of gratitude to Professor C. N. R. Rao, F. R. S., for introducing me to the fascinating world of monograph editing.

Kali Das Sen

Table of Contents

Similarity Models in the Theory of Pericyclic
Macromolecules
R. Ponec 1

Mapping of Molecular Electric Potentials and Fields
P. C. Mishra, and A. Kumar 27

Using Molecular Electrostatic Potential Maps
for Similarity Studies
G. Tasi, and I. Pálinkó 45

The Use of Graph Theoretical Methods
for the Comparison of the Structures of Biological
Macromolecules
P. J. Artymiuk, A. R. Poirette, D. W. Rice, and P. Willett 73

Identification of Structural Similarity
of Organic Molecules
Y. Takahashi 105

Author Index Volumes 151 - 174 135

Table of Contents of Volume 173

Molecular Similarity I

Similarity in Chemistry: Past, Present and Future
D. H. Rouvray

**Foundations and Recent Developments on Molecular
Quantum Similarity**
E. Besalú, R. Carbó, J. Mestres, M. Solà

**Density Domain Bonding Topology and Molecular
Similarity Measures**
P. G. Mezey

**Momentum-Space Electron Densities and Quantum
Molecular Similarity**
N. L. Allan, D. L. Cooper

Similarity Models in the Theory of Pericyclic Reactions

Robert Ponec

Institute of Chemical Process Fundamentals, Academy of Sciences of Czech Republic, Suchdol 2, Rozvojova 135, 16502 Prague 6, Czech Republic

Table of Contents

1 Similarity in Chemistry.   2
  1.1 Similarity in the Theory of Pericyclic Reactions.   2

2 Mechanisms of Pericyclic Reactions   6
  2.1 Concertedness in Pericyclic Reactivity   6
  2.2 More O'Ferrall Diagrams.   7
  2.3 Topological Criterion of Concertedness   9
  2.4 One-step vs Stepwise Mechanisms of Pericyclic Reactions. 13
  2.5 Least Motion Principle and the Mechanisms
      of Pericyclic Reactions. 18

3 Summary. 24

4 References 24

The recently introduced similarity approach to chemical reactivity is reviewed. The examples demonstrating the broad possibilities of the so-called similarity indices in the rationalisation of various mechanistic problems of pericyclic reactivity are discussed. The topics covered include (i) the alternative reproduction of the Woodward–Hoffmann rules, (ii) the formulation of a new topological criterion of concertedness, (iii) the formulation of a simple criterion discriminating between concerted one-step and nonconcerted stepwise mechanisms of pericyclic reactions and (iv) the variational formulation of the so-called least-motion principle as a new alternative means for analysing mechanisms of pericyclic reactions.

Topics in Current Chemistry, Vol. 174
© Springer-Verlag Berlin Heidelberg 1995

1 Similarity in Chemistry

There is probably no other concept that contributed to the development of chemistry so remarkably as the ill-defined, qualitative concept of similarity. Not despite but rather because of a certain fuzziness, the applicability of this concept is extremely broad and touches practically all areas of chemistry [1, 2]. An example would be the Mendeleev periodic law, the disclosure of which was closely connected with the effort to systematise the similarities in the properties of elements. From the intuitively interpreted meaning of similarity arises also one of the most powerful chemical principles – the principle of analogy, on the basis of which a wealth of fundamental chemical notions were introduced.

There are, for example, the concepts of isomerism, isoelectronicity, aromaticity, etc., but the interested reader can find a much more exhaustive account in the recent reviews by Rouvray [1, 2]. Although until quite recently the use of similarity concepts in chemistry still relied on a rather empirical intuitive basis, the rapid development of a new chemical discipline – mathematical chemistry – stimulated in recent years the numerous attempts to systematise the investigation of the similarity concept, especially from the point of view of the design of new quantitative measures of molecular similarity. Again because of fuzziness of the intuitively understood concept of similarity, the scope of these studies is extremely heterogeneous [4–24] and ranges from the attempts to characterise molecular shape [4–9] (important for the study of receptor–substrate interactions) to the application of various distance and similarity measures in the field of computer-aided synthesis [10–15] and in the design of compounds with desired chemical and biological properties [16–24]. Parallel to these efforts, there is also another group of more theoretically oriented papers attempting to relate the similarity in molecular properties to the similarity in the electron structure of the molecules [25–41]. The important role in this effort belongs to the so-called similarity indices, whose systematic exploitation is a subject of rapidly growing interest [23–41].

The increasing use of similarity ideas in recent years was enriched by a new and interesting attempt to extend the applicability of the approach beyond the scope of static structure–activity relationships to what is the heart of chemistry – chemical reactivity.

1.1 Similarity Ideas in the Theory of Pericyclic Reactions

The idea of the quantitative exploitation of molecular similarity for the classification of chemical reactivity is probably due to Polansky [27], who attempted to justify the old Clar's idea of local benzenoid regions in condensed aromatic hydrocarbons [42]. For this purpose, Polansky introduced a coefficient r_L characterising the similarity of a given local benzenoid fragment L in a molecule to the standard isolated benzene in terms of corresponding density matrices P and P_L (Eq. 1)

$$r_L = \frac{1}{2N_L} Tr P P_L \tag{1}$$

Because of normalisation by the factor $2N_L$ (N_L is the number of electrons in a fragment L), the values of the coefficient vary between 0–1, measuring thus the degree of similarity.

More recently the idea of quantitative comparison of electron structure of the molecules was revived by Carbo [28] who proposed for this purpose the similarity index r_{AB}, expressing the desired similarity of two molecules A and B in terms of density matrices (Eq. 2)

$$r_{AB} = \frac{\int \rho_A(1)\rho_B(1)\mathrm{d}\tau_1}{(\int \rho_A^2(1)\mathrm{d}\tau_1)^{1/2}(\int \rho_B^2(1)\mathrm{d}\tau_1)^{1/2}} \tag{2}$$

This definition also becomes the basis of our generalisation of the similarity index [33] directed for use in the theory of pericyclic reactivity. The first step of such a generalisation is to specify the molecules whose similarity we are interested in. In this respect, the natural choice is to identify the molecules A and B with the reactant and the product of a given reaction. The original index r_{AB} thus transforms into the index r_{RP}. The next important step of the generalisation is concerned with the approximations necessary for the practical calculation of the index. It appears that direct calculation of the index by straightforward integration according to Eq. (1) is not convenient. The most important drawback is concerned with the fact that the values of the index (1) are not invariant to the distance between and the mutual positions of the corresponding molecules. Analysis of this problem revealed that this inconvenient feature is generally caused by the presence of multicenter integrals $I_{\mu\nu\lambda\sigma}$ (Eq. 3).

$$I_{\mu\nu\lambda\sigma} = \int \chi_\mu^R \chi_\nu^R \chi_\lambda^P \chi_\sigma^P \mathrm{d}\tau \tag{3}$$

In order to eliminate the problems with the invariance, we proposed some time ago a topological approximation based on the so-called overlap determinant method [43]. This approximation is based on the transformation matrix T that describes the mutual phase relations of atomic orbitals centred on molecules R and P, and thus plays in this approach the same role as the so-called assigning tables in the overlap determinant method (Eq. 4)

$$\chi^P = T^R \chi \tag{4}$$

The construction of this matrix is sufficiently described in the original studies [33, 43] and therefore it is not necessary to repeat it here. We recall only that the form of the matrix depends on the actual reaction mechanism of the transformation $R \rightarrow P$ so that it is precisely via this matrix that the possibility of discriminating between various reaction mechanisms enters into play.

Combining Eq. (4) with the usual ZDO approximation for multicenter integrals $I_{\mu\nu\lambda\sigma}$ (Eq. 5) and the idempotency relations (Eq. 6),

$$I_{\mu\nu\lambda\sigma} \approx \delta_{\mu\lambda}\delta_{\nu\sigma} \tag{5}$$

$$\sum_{\mu}\sum_{\nu} p_{\mu\nu}^2 = 2N \tag{6}$$

the original definition equation (2) can be rewritten in the final form (Eq. 7), which can be regarded as a generalisation of the original formula by Polansky.

$$r_{RP} = \frac{1}{2N} Tr P_R \overline{P_P} \tag{7a}$$

$$\overline{P_P} = T^{-1} P_P T \tag{7b}$$

The most important feature of this generalisation consists in the direct inclusion of the transformation matrix T with which the various reaction paths can be quantitatively distinguished.

One of the above-mentioned possibilities of discriminating between various reaction paths is concerned with the alternative reproduction of the Woodward–Hoffmann rules [44]. This reproduction is based on a simple intuitive interpretation, according to which the magnitude of r_{RP} is a quantitative measure of the extent of electron reorganisation accompanying the transformation of the reactant into the product. Generally it is natural to expect that high values of r_{RP} can be regarded as indicating low electron reorganisation and vice versa. Such an intuitive interpretation is especially interesting because of its relation to the so-called least-motion principle [45] according to which the reactions prefer those reaction paths along which the reacting molecules undergo minimal changes in nuclear and electronic configurations. In view of the above relationship between the magnitude of r_{RP} and the extent of electronic reorganisation, the ease of the reaction can be characterised by the value of the similarity index.

This intuitive parallel can be best demonstrated by the example of electrocyclic reactions for which the values of the similarity indices for conrotatory and disrotatory reactions systematically differ in such a way that a higher index or, in other words, a lower electron reorganisation is observed for reactions which are allowed by the Woodward–Hoffmann rules. In contrast to electrocyclic reactions for which the parallel between the Woodward–Hoffmann rules and the least motion principle is entirely straightforward, the situation is more complex for cycloadditions and sigmatropic reactions where the values of similarity indices for alternative reaction mechanisms are equal so that the discrimination between allowed and forbidden reactions becomes impossible. The origin of this insufficiency was analysed in subsequent studies [46, 47] in which we demonstrated that the primary cause lies in the restricted information content of the index r_{RP}. In order to overcome this certain limitation, a solution was proposed based on the use of the so-called second-order similarity index g_{RP} [46]. This index, introduced in direct analogy with the first order index r_{RP} is generally defined by Eq. (8), where $\rho_R(1,2)$ and $\rho_P(1,2)$ are the second-order density matrices (pair densities) of the molecules R and P.

$$g_{RP} = \frac{\int \rho_R(1,2)\rho_P(1,2)d\tau_1 d\tau_2}{(\int \rho_R^2(1,2)d\tau_1 d\tau_2)^{1/2}(\int \rho_P^2(1,2)d\tau_1 d\tau_2)^{1/2}} \qquad (8)$$

Within the framework of the same topological approximation as for the index r_{RP}, this general formula can be rewritten in an alternative form used in actual calculations (Eq. 9).

$$g_{RP} = \frac{9Tr^2(P_R\overline{P_P}) - 7Tr(P_R\overline{P_P})^2}{4N(9N-14)} \qquad (9)$$

The most important difference resulting from the use of pair density matrices is that they include, even if only partially, electron correlation so that the increased information content of the index g_{RP} can be related precisely to these correlation contributions. A comparison of the values of first- and second-order similarity indices calculated from simple HMO wave functions is given in Table 1.

The most interesting result of such a comparison is that the original insufficiency of the first order index r_{RP} is indeed remedied by the index g_{RP}. This suggests that the role of electron correlation is probably higher in cycloadditions and sigmatropic reactions than in the electrocyclic ones. In this connection it is also interesting to mention the case of the Cope rearrangement for which discrimination between allowed and forbidden mechanisms is impossible with either r_{RP} and g_{RP}. This reaction was analysed in a separate study [47] in which we demonstrated that the correct discrimination requires, in this case, the introduction of higher than pair correlations in the so-called third-order similarity index.

Table 1. Comparison of first- and second-order similarity indices for several selected pericyclic reactions

Reaction	Mechanism[a]	r_{RP}	g_{RP}
Butadiene/cyclobutene	Conrotation	0.724	0.524
	Disrotation	0.500	0.091
Hexatriene/cyclohexadiene	Disrotation	0.759	0.566
	Conrotation	0.659	0.358
Ethene dimerisation	$s + a$	0.500	0.250
	$s + s$	0.500	0.091
Diels-Alder reaction	$s + s$	0.575	0.298
	$s + a$	0.575	0.272
Cope rearrangement	$s + s$	0.500	0.206
	$s + a$	0.500	0.206

[a] The upper value corresponds to an allowed and the lower value to a forbidden reaction mechanism

Such an alternative reproduction of the Woodward–Hoffmann rules is not, however, the only result of the similarity approach. The formalism of the approach is very flexible and universal and the original study [33] has become the basis of a number of subsequent generalisations in which a number of problems dealing with various aspects of pericyclic reactivity were analysed and discussed [48–55]. In spite of the considerably broad scope of these applications, recently reviewed in [56], the possibilities of the similarity approach are still not exhausted and its formalism is still capable of further methodological development. Our aim in this report is to present some of more recent applications of the similarity approach for the study of mechanisms of pericyclic reactions [57–59].

2 Mechanisms of Pericyclic Reactions

The concept of reaction mechanism is very broad and its exact meaning depends to considerable extent on the point of view from which a given problem is to be analysed. Thus, for example, reaction mechanisms can be understood differently by a chemical physicist analysing a given reaction at the level of elementary collisions in crossed molecular beams, and by an organic chemist analysing the reaction course by the formalism of phenomenological kinetics. This implies that if one wants to speak about the mechanism of the reaction it is always necessary to specify also the point of view, from which the reaction is analysed. Thus, for example, in the case of usual reactions performed on the preparative scale, the term reaction mechanism is used to denote the detailed specification of whether the reaction proceeds in one elementary step or whether some, more or less stable, intermediates intervene.

2.1 Concertedness in Pericyclic Reactivity

Although such an understanding of the reaction mechanism is in principle applied in the theory of pericyclic reactions, the above general picture is in this case slightly complicated by the specific (introduced in the course of historical development) classification of reaction mechanisms in terms of concertedness and/or nonconcertedness. Concerted reactions are intuitively understood as those reactions for which the scission of old bonds and the formation of the new ones is synchronised, whereas for nonconcerted reactions the above bond exchange processes are completely asynchronised. Moreover, since the above asynchronicity is also intuitively expected to induce the stepwise nature of the process, the nonconcertedness is frequently believed to require the presence of intermediates, whereas the concerted reactions are believed to proceed in one elementary step.

Although such a definition is seemingly quite clear and unique, the practical exploitation of the above criterion is complicated by the fact that the scission and formation of bonds is a microscopic process, inaccessible to direct experimental observation. This, of course, suggests the necessity of searching other, more easily exploitable, criteria of concert. One such criterion is the remarkable stereospecificity accompanying the formation of products in allowed pericyclic reactions [60, 61]. The fact that the origin of the synchronisation in the process of scission and the formation of the bonds was always intuitively related to a certain energetic stabilisation led to another widespread opinion that all allowed reactions are automatically concerted. On the other hand nonconcertedness, advocated by frequently observed stereo-randomization [60] was practically always expected in forbidden reactions.

The continuing accumulation of experimental material made it, however, still more apparent that the above criteria, both kinetic and stereochemical [61–64], cannot be regarded as absolute [65], since there are probably both concerted reactions which are forbidden as well as nonconcerted ones which are allowed. An example of the first type of reaction is some $2 + 2$ retro-cycloadditions on sterically strained systems [62, 66]; the second type of processes is then represented by the so-called multibond reactions [67]. This, of course, stimulates to formulate a new and more universal criterion of concertedness as well as a deeper understanding of the very concept of concertedness.

One of the most fruitful in this respect is the study [68], in which it was demonstrated that a certain confusion concerning the concept of concertedness is apparently due to the fact that this concept has two different aspects, structural and energetic, which are not always properly distinguished. The structural aspect operates with the concept of synchronicity and/or asynchronicity. On the other hand, from the energetic point of view, the primary role is played by the form of the reaction profile. The cause of the above-mentioned confusion is then the fact that the term concertedness is frequently used as a synonym for the synchronisation of bonding changes and also to denote a process for which the reaction profile has no secondary minima corresponding to intermediates. This duality suggests that a solution to the problem of concertedness depends upon a precise definition of its meaning. The important contribution in this respect is represented by the technique of the so-called More O'Ferrall diagrams [69, 70], on the basis of which it was possible to formulate a new universal problems concerning the duality of the concept of concertedness, we consider it convenient, before starting the presentation of the new criterion, to recapitulate briefly the basic ideas of the original technique of the More O'Ferrall diagrams.

2.2 More O'Ferrall Diagrams

The basic idea of this technique can be best demonstrated by reactions whose is governed by the variation of two nuclear coordinates Q_1 and Q_2, representing, for example, the lengths of two newly created bonds. An example of such a process would be the Diels-Alder reaction (Scheme 1).

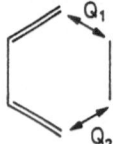

Scheme 1.

If we now regard the reaction mechanism classically as a sequence of consecutive scissions and formation of bonds, the course of the reaction can be described by a line (corresponding to reaction path) connecting the reactant and the product in a schematic diagram, in which the degree of formation of individual bonds is evaluated on individual axes by the conventional scale ranging from 0 to 1 (Fig. 1). Apparently, the concerted process, characterised by ideal synchronisation in the formation of both bonds is, within the framework of this approach, depicted by the diagonal line connecting the opposite corners R and P. The second limiting case then corresponds to the two-step line going along the periphery and including the corners with the coordinates $[0, 1]$ and $[1, 0]$, corresponding to intermediates.

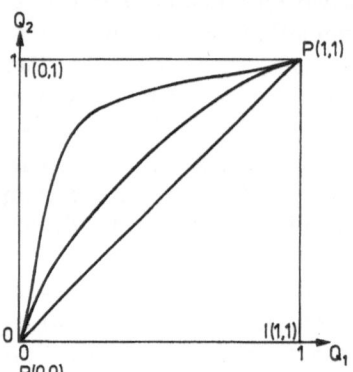

Fig. 1. Schematic representation of reaction paths in the More O'Ferrall diagram

Between these two idealised extremes there then exists a practically continuous scale of cases corresponding to the majority of real situations, where the asynchronisation in the formation of bonds is only partial. Since the degree of this asynchronisation is given by the detailed form of the corresponding line, it is apparent that the extent of the deviations from the ideally synchronous diagonal line or the closeness of the approach to the ideally nonconcerted two-step line provides a simple means of characterising the concertedness and/or nonconcertedness of the reaction. In spite of its conceptual simplicity, the above criterion is, however, difficult to apply practically since the inevitable arbitrariness in evaluating the degree of the corresponding deviations makes it difficult to decide where precisely, on a practically continuous scale of processes differing in the degree of bonding synchronisation, lies the borderline separating the asynchronous but still concerted processes from the nonconcerted ones. As will be shown below, by appropriate generalisation of the original More O'Ferrall diagrams, precisely this arbitrariness can be eliminated.

2.3 Topological Criterion of Concertedness

The basis of this criterion, representing another example of the exploitation of similarity indices, is the discovery that the set of structures lying "inside" the corresponding More O'Ferrall diagrams can be dissected into certain regions, the characteristic feature of which is that each of them surrounds one of the key molecular species participating in the reaction, i.e. the reactant, product and intermediate [57]. In this connection, it is interesting that the possibility of dissection into certain so-called catchment regions was reported some time ago by Mezey [71] for potential energy surfaces. Our regions of dominant similarity can thus be intuitively regarded as a certain analogy of Mezey's catchment regions. On the basis of the above dissection, the criterion of concertedness can be formulated quite straightforwardly according to which regions are actually crossed by a given reaction path. The situation can be best visualised graphically (Fig. 2) in a schematic diagram, on the basis of which is quite natural to classify as concerted the process for which the reaction path goes, as in the case of an ideally synchronous process, from the region of the reactant directly into the region of the product. On the other hand, where the reaction path enters into the region of the intermediate, the corresponding reaction can be characterised as nonconcerted.

In connection with this criterion, it is of course apparent that its practical exploitation is conditioned by two fundamental factors. One of them is the

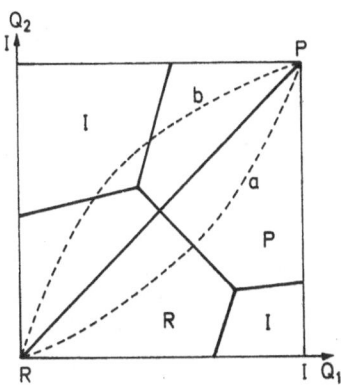

Fig. 2. Schematic classification of reaction mechanisms in terms of partitioning of modified More O'Ferrall diagrams. *Line a* represents an asynchronous but concerted process, *line b* a nonconcerted process

knowledge of the corresponding reaction path and the other is the question of the detailed procedure allowing the dissection of the diagrams into individual regions. Both these questions will be discussed in detail below. Here we start with the second problem, which is the dissection of the More O'Ferrall diagram into the regions. The philosophy of this dissection can be best demonstrated by the simplest case of a process in which the reactant R is transformed into the product P in one elementary step. Such a process can be described, in the sense of a generalised overlap determinant method [72], by the parametrical trans-formation equation (10)

$$\Psi(\varphi) = \frac{1}{N(\varphi)}(\Psi_R \cos\varphi + \Psi_P \sin\varphi) \tag{10}$$

in which the parameter φ, playing the role of the generalised reaction coordinate varies within the range $(0, \pi/2)$ for allowed reactions and within $(0, -\pi/2)$ for forbidden ones [50]. On the basis of this equation, the so-called generalised similarity indices $r_{RX(\varphi)}$, $r_{PX(\varphi)}$ (Eq. 11) were introduced [48–49], characterising the similarity of the general transient species $X(\varphi)$ to the starting reactant or the final product in the general point of the concerted reaction path.

$$r_{RX(\varphi)} = \frac{1}{N(\varphi)} Tr P_R \overline{P(\varphi)} \tag{11}$$

If we now plot the dependencies $r_{RX(\varphi)}$, $r_{PX(\varphi)}$ vs $\varphi(r_{RX(\varphi)} = r_{PX(\pi/2-\varphi)})$, then the result is a picture of two mutually crossing curves. One of these curves describes the variation of the electron reorganisation in the process $R \rightarrow P$, whereas the other describes the same dependence for the backward process $P \rightarrow R$ (Fig. 3).

As can be seen in Fig. 3, both curves cross at the point $X(\pi/4)$, satisfying the interesting relation (12) and expressing the fact that the similarity of this critical point both to the reactant and the product is the same.

$$r_{RX(\pi/4)} = r_{PX(\pi/4)} \tag{12}$$

This critical point thus bisects the whole reaction path into two disjunct regions in such a way that the elements of one class are structurally closer (more similar)

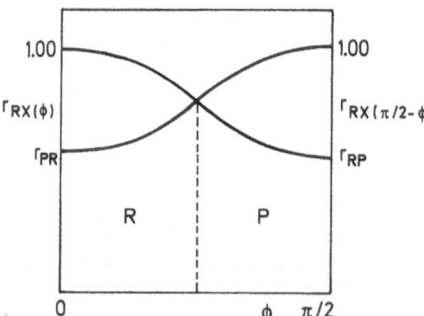

Fig. 3. Schematic visualisation of the partitioning of a concerted reaction path to the region of the reactant and the region of the product

to the reactant than to the product whereas for the second class the similarity to the product dominates. The criterion of this dissection is then simply given by the set of inequalities (13).

$$r_{RX(\varphi)} > r_{PX(\varphi)} \quad X \in R \tag{13a}$$

$$r_{PX(\varphi)} > r_{RX(\varphi)} \quad X \in P \tag{13b}$$

The same philosophy which in the case of one-dimensional model led to the introduction of the regions of the reactant and the product can be applied also to more complex reaction schemes involving eventual intermediates. The simplest situation is in the case of More O'Ferrall diagrams where the participating intermediate is only one. The corresponding generalisation is very simple and consists in the straightforward modification of the original Eq. (10) into the form (14) explicitly involving the intermediate via the approximate wave function Ψ_I.

$$\Psi(\vartheta, \varphi) = \frac{1}{N(\vartheta, \varphi)} (\Psi_R \cos \vartheta \cos \varphi + \Psi_P \cos \vartheta \sin \varphi + \Psi_I \sin \vartheta) \tag{14}$$

Similarly, as with Eq. (10), the parameters ϑ and φ play again the role of generalised reaction coordinates, the systematic change of which allows the structure of all transient species lying "inside" the corresponding More O'Ferrall diagrams to be described. In order that the analogy with these diagrams becomes even more transparent, it is convenient to substitute the primary spherical coordinates ϑ and φ by ordinary More O'Ferrall coordinates Q_1 and Q_2. These transformations are described by the set of equations (15).

$$Q_1 = \frac{1}{\pi}\left(2\varphi + |\vartheta|\left(1 - \frac{4\varphi}{\pi}\right) - \vartheta\right) \tag{15a}$$

$$Q_2 = \frac{1}{\pi}\left(2\varphi + |\vartheta|\left(1 - \frac{4\varphi}{\pi}\right) + \vartheta\right) \tag{15b}$$

On the basis of these transformation relations, it is then possible to introduce the topological density matrix $P(Q_1, Q_2)$ (Eq. 16), and consequently, also the individual similarity indices $r_{RX(\vartheta, \varphi)}, r_{PX(\vartheta, \varphi)}, r_{IX(\vartheta, \varphi)}$.

$$P(Q_1, Q_2) = N \int \Psi^2(Q_1, Q_2) d\xi_1 dx_2 dx_3 \ldots dx_N \tag{16}$$

The corresponding dissection of the More O'Ferrall diagram is then given by the set of inequalities (17):

$$r_{PX(Q_1, Q_2)} < r_{RX(Q_1, Q_2)} > r_{IX(Q_1, Q_2)} \quad X \in R \tag{17a}$$

$$r_{RX(Q_1, Q_2)} < r_{PX(Q_1, Q_2)} > r_{IX(Q_1, Q_2)} \quad X \in P \tag{17b}$$

$$r_{RX(Q_1, Q_2)} < r_{IX(Q_1, Q_2)} > r_{PX(Q_1, Q_2)} \quad X \in I \tag{17c}$$

In connection with this conceptually very simple criterion it is also useful to notice that its credibility is supported not only intuitively but can also be justified theoretically in a way which enlightens the origins of the frequently expected, but also criticised parallelism between the synchronisation of the process of bonding reorganisation and the shape of the energy profile for the reaction. Such theoretical support originates from the study [73], dealing with the mutual relation between the structural and energetic stability of the system. The conclusions of this study can be briefly summarised by saying that the energetic stability of the system implies also its structural stability. From this then follows, in a reversed implication, that a system that is unstable structurally will also be unstable energetically. Since the important points (or regions) of structural instability are precisely those borderlines that separate in our approach the individual species R, P and I, it is apparent that the points where the reaction coordinate crosses the border separating two structurally stable regions in the More O'Ferrall diagram must correspond, on the basis of the above theorem, to the energetically unstable point on the PE hypersurface. This, together with the results of the study by Mezey [71] implies that any nonconcerted reaction, i.e. a reaction going according to our criterion via two borderlines of structural instability, necessarily proceeds by a two-step reaction mechanism involving at least one energetically stable intermediate. This straightforward mutual parallel to the generally accepted interpretation that regards as synonymous the concepts of a nonconcerted reaction and a reaction going via intermediates cannot be, unfortunately, transferred to concerted reactions since the fact that in the concerted process the reaction path passes only one borderline of structural instability does not imply that the energetic profile of the reaction necessarily needs to go (even if it can) via only one energetic maximum. Just here, in the non-equivalence of the concepts of structural and energetic instability, is thus apparently hidden the origin of the problems arising from the widespread, but incorrect identification of concertedness as a synonym for a reaction without intermediates [68].

The practical exploitation of the proposed criterion can be very simply demonstrated by the example of the electrocyclic transformation of cyclobuta-diene to cyclobutene, for which the structure of the intermediate can be quite reliably estimated from the available results of quantum chemical calculations [74] (Scheme 2).

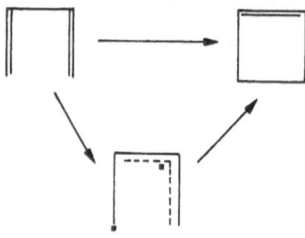

Scheme 2.

This reaction is especially convenient for demonstration purposes since it displays both possible ways in which the More O'Ferrall diagrams can be dissected, as schematically depicted in Figs. 4 and 5.

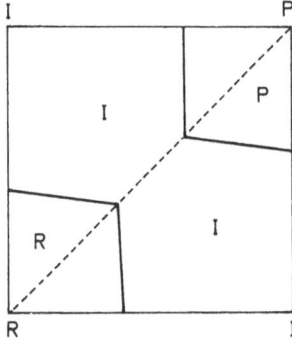

Fig. 4. Schematic visualisation of the specific partitioning of the More O'Ferrall diagram in the case of a reaction allowing classification of the reaction mechanism without a knowledge of the actual reaction path

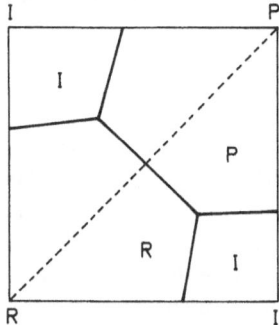

Fig. 5. Schematic visualisation of the most usual form of the More O'Ferrall diagram requiring a knowledge of the actual reaction path for classifying the reaction mechanism

Especially interesting is the case of forbidden disrotatory cyclisation, for which the special form of the dissection allows classification of the mechanism even without knowledge of the reaction path. As can be seen from the Fig. (4), no reaction path connecting the reactant and the product can avoid the region of the intermediate so that the reaction has to be classified as nonconcerted. Such cases are not, however, very frequent and in the majority of cases the topology of the dissection is such that without knowledge of the actual reaction path the classification of the reaction mechanism is impossible (Fig. 5).

Before starting a detailed presentation of the procedure according to which the reaction paths can be determined, we show first a simpler but nevertheless useful criterion, enabling us to obtain the first rough idea of the relative ease of concerted and nonconcerted reaction paths.

2.4 One-Step vs Stepwise Reaction Mechanisms of Pericyclic Reactions

The basic idea of the above criterion is very simple and consists in the qualitative comparison of the relative ease of selected idealised reaction paths corresponding in More O'Ferrall diagrams to purely synchronous (concerted) and purely asynchronous (stepwise) reaction paths (Eq. 18).

$$R \to P \tag{18a}$$

$$R \to I \to P \tag{18b}$$

In the case of concerted reaction paths (18a) the ease of the reaction can be characterised, as shown above in Sect. 1.1 by the value of the similarity index r_{RP}, or, better by the difference $(1 - r_{RP})$ [53]. The question thus only is how to generalise the similarity approach to the evaluation of the ease of the stepwise reaction paths (18b). For this purpose we proposed some time ago [58] a simple model characterising the stepwise reaction paths in terms of the similarity indices r_{RI} and r_{IP}, corresponding to individual reaction steps. In terms of these indices, the ease of the stepwise reaction mechanism can be characterised by the quantity L (Eq. 19).

$$L_{\text{stepwise}} = (1 - r_{RI}) + r_{RI}(1 - r_{IP}) = (1 - r_{RP}) \tag{19}$$

The derivation of this expression is based on the following idea. Let us suppose for a moment that the reaction steps (18b) are completely independent. In this case the total extent of electron reorganisation would simply be the sum of the terms $(1 - r_{RI})$ and $(1 - r_{IP})$ which characterise the electron reorganisation in individual reaction steps. Such is not, however, the case, since the existence of a common intermediate that is simultaneously the product of the first and the reactant of the second step, makes the reactions mutually coupled. In order to take this coupling into account, it is necessary to modify the expression for the extent of the electron reorganisation in the second step. This modification appears as the multiplicative term r_{RI} in front of the second member of Eq. (19). Within the framework of this simple model, the relative ease of concerted and stepwise reaction mechanisms can be estimated by comparing the similarity index r_{RP} with the product $r_{RI}r_{IP}$. Where the value of r_{RP} is larger than the product $r_{RI}r_{IP}$, the concerted mechanism can be expected to be preferred whereas the greater value of the product allows one to expect rather the preference of stepwise mechanism.

As an example of the practical use of the above criterion, let us discuss again the electrocyclic transformation of butadiene to cyclobutene. The individual alternative reaction mechanisms are described, as in the previous chapter, by the scheme II. For evaluating relative ease of individual reaction paths, it is necessary to first calculate the density matrices P_R, P'_I and P''_P which are then, in the next step converted into the similarity indices r_{RP}, r_{RI} and r_{IP}. Such a calculation requires, however, the density matrices to be transformed into the common basis of atomic orbitals [33, 43]. These transformations are described by the matrices T_{RP}, T_{RI} and T_{IP} which has to be determined for each elementary step.

$$R \xrightarrow{\;T_{RI}\;} I \xrightarrow{\;T_{IP}\;} P \qquad R \xrightarrow{\;T_{RP}\;} P \tag{20}$$

Since the detailed calculation of these matrices is sufficiently described in the original literature [33, 58], it is possible to present directly the final results first for the case of concerted reactions for which there are two alternative reaction mechanisms, conrotatory and disrotatory. The first of these mechanisms is allowed by the Woodward–Hoffmann rules while the second one is forbidden.

$$r_{RP}^{con} = 0.724$$

$$r_{RP}^{dis} = 0.500 \tag{21}$$

If we now look at the values of the above indices, it is possible to see that the prediction of the Woodward–Hoffmann rules is indeed confirmed since the greater values of the similarity index for the conrotatory reaction clearly imply, in keeping with the expectations of the least-motion principle, the lower electron reorganisation. If now the same formalism is applied to a stepwise reaction mechanism, the following values of the similarity indices result (Eq. 21).

$$r_{RI}^{con} = r_{RI}^{dis} = 0.851$$

$$r_{IP}^{con} = r_{IP}^{dis} = 0.781 \tag{22}$$

These values are quite interesting since, as can be seen, there is no difference in this case between conrotatory and disrotatory reaction mechanisms. This result, at first sight surprising, is, however, completely reasonable since it in fact reflects nothing but the lack of stereospecificity generally expected for non-concerted reaction paths. More interesting than this first qualitative result however, is, the comparison of the relative ease of concerted and stepwise reaction mechanisms. Here it is possible to see that whereas in the case of an allowed conrotatory reaction the concerted mechanism seems to be preferred $(r_{RP} > r_{RI}r_{IP})$, in the case of a forbidden disrotatory reaction it is the stepwise nonconcerted reaction mechanism which is more favourable $(r_{RI}r_{IP} > r_{RP})$. This result is very interesting since just the same conclusions can be expected in this case not only on the basis of intuitive considerations deriving from the Woodward–Hoffmann rules but also for corresponding remarkably with what is predicted for this reaction from available theoretical data [74, 75–77]. Especially interesting in this connection is the case of a forbidden disrotatory reaction for which both semiempirical and ab initio calculations predict the reaction path to be very asynchronous. The limiting case of this asynchronous path is then our idealised stepwise mechanism.

In addition to this mechanistically most valuable result, some other conclusions can also be deduced by comparing the individual similarity indices r_{RI} and r_{IP} which characterise the extent of electron reorganisation in individual reaction steps. In our case it is possible to see that the extent of electron reorganisation is greater in the second step $(r_{RI} > r_{IP})$. This implies that this step should also be rate determining. The reaction profile of such a process thus corresponds to what Dewar calls the two-step process [78]. It would be certainly interesting to confront this theoretical prediction with experiment but unfortunately the lack of convenient data makes such a comparison unfeasible. As we shall see below, however, such a comparison is possible in other cases and some of these will be discussed later. The calculated values of the similarity indices for the set of selected pericyclic reactions are summarised in Table 2.

Table 2. Calculated values of similarity indices r_{RP}, r_{RI} and r_{IP} for several selected pericyclic reactions

Reaction	Mechanism[a]	r_{RP}	r_{RI}	r_{IP}
Butadiene/cyclobutene	Conrotation	0.724	0.851	0.781
	Disrotation	0.500	0.851	0.781
Hexatriene/cyclohexadiene	Disrotation	0.759	0.895	0.835
	Conrotation	0.659	0.895	0.835
Ethene dimerisation	$s + a$	0.500	0.577	0.866
	$s + s$	0.500	0.577	0.866
Diels-Alder reaction	$s + s$	0.575	0.721	0.859
	$s + a$	0.575	0.721	0.859
Cope rearrangement	$s + s$	0.500	0.730	0.730
via 1,4-cyclohexadiyl	$s + a$	0.500	0.730	0.730

[a] The upper value corresponds to an allowed and the lower to a forbidden reaction mechanism

We discuss now some general features arising from the above table. The simplest situation is in the case of electrocyclic transformation of hexatriene to cyclohexadiene where the calculated values strongly suggest the preference of a concerted mechanism for the allowed disrotation while for the forbidden conrotation the favourable reaction path should be the stepwise one. All these prediction are again in complete harmony with the expectations of the Woodward–Hoffmann rules. A rather more complex situation is found in the case of cycloadditions and sigmatropic reactions, the specific position of which was already mentioned. The most delicate situation s found in the case of $2 + 2$ ethene dimerisation where the equality of r_{RP} for both $s + s$ and $s + a$ mechanisms as well as of the product $r_{RI}r_{IP}$ suggests that any discrimination between alternative reaction mechanisms is impossible in this case. Fortunately there is a possibility of overcoming this drawback. As in the case of discrimination between allowed and forbidden concerted mechanisms, this possibility us based on the use of the second order similarity index g. If now the same remedy is applied to the discrimination between one-step and stepwise mechanisms, the original criterion based on the comparison of r_{RP} with the product $r_{RI}r_{IP}$ analogously requires a comparison of g_{RP} with the product $g_{RI}g_{IP}$. As with first order indices, the concerted one-step mechanism can be expected to be preferred for $g_{RP} > g_{RI}g_{IP}$ whereas for $g_{RP} < g_{RI}g_{IP}$ the stepwise mechanism should be favoured. The calculated values of second order similarity indices for the same set of selected pericyclic reactions are summarised in Table 3.

As can be seen from the table, the values of g_{RP} do indeed remedy the insufficiency of the first order index r and correctly predict the reaction preferred

Table 3. Calculated values of second order similarity indices g_{RP}, g_{RI} and g_{IP} for several selected pericyclic (multibond) reactions

Reaction	Mechanism[a]	g_{RP}	g_{RI}	g_{IP}
Ethene dimerisation	$s + a$	0.250	0.241	0.724
	$s + s$	0.091	0.241	0.724
Diels-Alder reaction	$s + s$	0.298	0.468	0.711
	$s + a$	0.272	0.468	0.711
Cope rearrangement	$s + s$	0.206	0.481	0.481
via 1,4-cyclohexadiyl	$s + a$	0.206	0.481	0.481

[a] The upper value corresponds to an allowed and lower to a forbidden reaction mechanism

for the concerted process by the Woodward–Hoffmann rules. Much more interesting than this reproduction of the Woodward–Hoffmann rules, however, is the comparison between the relative ease of concerted and stepwise reaction mechanisms. Thus, for example, in the case of 2 + 2 ethene dimerisation where the original criterion failed, the new enhanced criterion based on second order indices clearly suggests that the preferred mechanism of forbidden $s + s$ dimerisation is the stepwise one. This is in complete agreement with both experimental and theoretical data for this process [79, 80].

In addition to this prediction of the preferred reaction mechanism, the individual values of the similarity indices g_{RI} and g_{IP} (as well as the values of r_{RI} and r_{IP}) which provide a measure of the extent of electron reorganisation in individual steps $R \to P$ and $I \to P$ can be used to estimate which of the reaction steps can be expected to determine the rate. As can be seen from both Tables 3 and 4, the rate-determining step for $s + s$ stepwise ethene dimerisation should be the formation of the intermediate. Although the lack of appropriate experimental data prevents direct verification of this theoretical prediction, it is interesting that this prediction is in complete agreement with available theoretical calculations [79, 80].

In a completely analogous way it should now be possible to analyse any other multibond reaction. Especially interesting in this respect is the Diels-Alder reaction, for which the question of the eventual participation of a nonconcerted stepwise mechanism was suggested some time ago by Dewar [81, 82]. As can be seen from the Tables 3 and 4, comparison between concerted and stepwise mechanisms seems to suggest, irrespective of whether first or second order indices are used, a preference for the stepwise mechanism, the first step being rate determining. This corresponds to a reaction of the two-stage type [78]. In this connection it is interesting that the same type of mechanism, combining the observed kinetic concertedness with the possibility of certain asynchronisation in the formation of the bonds was proposed for the Diels-Alder reaction some time ago by Woodward and Katz [83]. The preference for an asynchronous stepwise mechanism is in this case also in harmony with the results of semiempirical calculations by Dewar [81, 82]. Since however, these results were questioned by more sophisticated ab initio techniques [75, 84–87] and since the parent unsubstituted systems behave frequently atypically [84], we will make our prediction in this case with caution. Nevertheless it is interesting that a similar prediction of a two-stage reaction mechanism in the case of closely related 2 + 4 dimerisation of butadiene (Scheme 3) is in complete harmony with experimental results by Doering [88] and also with very recent ab initio calculations by Houk [89].

Scheme 3.

2.5 Least Motion Principle and the Mechanisms of Pericyclic Reactions

Although the above criterion, allowing as it does a simple estimate of the relative ease of concerted and nonconcerted reaction paths, is certainly interesting and useful, a detailed analysis of reaction mechanisms in terms of More O'Ferrall diagrams clearly demonstrated that the idealised picture of the reaction mechanism, that takes into account only ideally synchronous and ideally nonconcerted reaction paths, is only very crude and simplistic. The above two reaction mechanisms correspond, namely, only to certain limiting situations, whereas in the majority of real cases the reaction proceeds in such a way that the synchronisation in the scission and formation of bonds is only partial. Typical and quite frequent examples in this respect are various elimination reactions, for which the intermediate reaction mechanism often lies between the E1 and E2 extremes [90]. Such a concept of a reaction path as a trajectory describing the detailed synchronisation in the process of scission and the formation of bonds is also the basis of the generalisation of the least motion principle. This generalisation is based on the introduction of appropriate quantities that allow the extent of electron reorganisation along an arbitrary reaction path to be characterised. On the basis of these quantities, the reaction path satisfying the least motion principle can be determined variationally from the condition of minimal electron reorganisation.

Prior to a detailed presentation of the criterion that allows the corresponding reaction paths to be determined, it is convenient to mention also some conceptual difference that differentiate our reaction paths from the usual quantum chemical interpretation, based on the concept of potential energy hypersurfaces. The most striking difference concerns that fact that in contrast to usual quantum chemical interpretation, based on the concept of potential energy hypersurfaces. The most striking difference concerns the fact that in contrast to usual quantum chemical model of IRC as a trajectory in the configuration space of the nuclei, our model of the reaction path employs a classical chemical picture of the reaction mechanism as a sequence of electron shifts, depicting the gradual scission and formation of individual bonds. For this reason it would be probably difficult to make any straightforward comparisons of both approaches; and, even if we believe that some parallels do exist, their disclosure is a matter of future systematic investigations.

Based on the introductory presentation of the philosophy of the approach let us proceed now to the discussion of the mathematical part of the formalism [59]. The basis of the model is the topological model [57] described by the generalised wave function (22),

$$\Psi(\vartheta, \varphi) = \frac{1}{N(\vartheta, \varphi)} (\Psi_R \cos\vartheta \cos\varphi + \Psi_P \cos\vartheta \sin\varphi + \Psi_I \sin\vartheta) \qquad (22)$$

where the possible asynchronicity of the reaction path is given by the parameter $\vartheta = \vartheta(\varphi)$. Within the framework of this model, the knowledge of the dependence $\vartheta = \vartheta(\varphi)$ is equivalent to the knowledge of the reaction mechanism and the aim of the theoretical calculations is to find the form of this dependence.

In connection with Eq. (22), yet another important factor differentiates our approach from usual quantum chemical analyses of reaction mechanisms. This difference concerns the fact that while a quantum chemical approach is in principle independent of any external information (all participating species appear automatically as various critical points on the PE hypersurface), in our model that is more closely related to classical chemical ideas some auxiliary information about the structure of the participating molecular species is required. This usually represents no problem with the reactants and the products since their structure is normally known, but certain complications may appear in the case of intermediates. This complication is not, however, too serious since in many cases the structure of the intermediate can be reasonably estimated either from some experimental or theoretical data or on the basis of chemical intuition. Thus, for example, in the case of pericyclic reactions that are of primary concern for us here, the intermediates are generally believed to correspond to biradical or biradicaloid species with the eventual contributions of zwitterionic structures in polar cases.

Assuming that in a given case the structures of all the participating molecular species is known, it is possible to begin the practical exploitation of Eq. (22) and aim at the variational formulation of the least-motion principle. For this purpose, it is first necessary to introduce the first order density matrix $\rho(\vartheta, \varphi|x_1)$, (Eq. 23), where x_1 denotes the position vector of the i-th electron, ξ_1 its spin coordinate and N the total number of electrons

$$\rho(\vartheta, \varphi|x_1) = N \int \Psi^2(\vartheta, \varphi) d\xi_1 dx_2 \ldots dx_N \qquad (23)$$

Using this matrix, the difference in the electron configurations of two different structures corresponding to different values of the parameters ϑ and φ (e.g. ϑ, φ

and ϑ', φ') can be quantitatively characterised by the similarity index K (Eq. 24)

$$K = \frac{\int \rho(\vartheta, \varphi|x_1)\rho(\vartheta', \varphi'|x_1)dx_1}{(\int \rho^2(\vartheta, \varphi|x_1)dx_1)^{1/2}(\int \rho^2(\vartheta', \varphi'|x_1)dx_1)^{1/2}} \tag{24}$$

Expressing the individual density matrices on the basis of atomic orbitals (Eq. 25) and using the same topological approximation as in the introductory study [33], the above general expression can be rewritten in the final form (26).

$$\rho(\vartheta, \varphi|x_1) = \sum_\mu \sum_\nu [P(\vartheta, \varphi)]_{\mu\nu}\chi_\mu(x_1)\chi_\nu(x_1) \tag{25}$$

$$K = \frac{TrP(\vartheta, \varphi)P(\vartheta', \varphi')}{[TrP^2(\vartheta, \varphi)]^{1/2}[TrP^2(\vartheta', \varphi')]^{1/2}} \tag{26}$$

As can be seen, this index attains its maximal value of unity for two identical structures ($\vartheta' = \vartheta$, $\varphi' = \varphi$) and with increasing deviations of both structures its value monotonously decreases. The use of this index for the formulation of the least motion principle arises from the following simple idea. Let us assume that we are on a reaction path at point characterised by the wave function $\Psi(\vartheta', \varphi')$ and we are looking for such an infinitesimally close structure $\Psi(\vartheta, \varphi)$ for which the transformation $\Psi(\vartheta', \varphi') \to \Psi(\vartheta, \varphi)$ requires minimal change in electronic configuration. This condition is equivalent to a search of the direction along which the derivative of K at the point $\Psi' = \vartheta$ and $\varphi' = \varphi$ attains its minimum. This directional derivative can be mathematically described as (27),

$$\frac{dK}{dl} = s_\vartheta \nabla_\vartheta K + s_\varphi \nabla_\varphi K \tag{27}$$

Direct calculation of the gradient of index K demonstrates, however, that all its components are identically zero. This implies that there is no direction along which the limiting first derivative of K is non-zero. This suggests that the characterisation of the transformation $\Psi(\vartheta', \varphi') \to \Psi(\vartheta, \varphi)$ is to be based on a second directional derivative. Taking now into account that the index K attains its maximum at the point $\vartheta' = \vartheta$ and $\varphi' = \varphi$, then the reaction path satisfying the requirement of minimal electron reorganisation should continue in the direction along which the second derivative is maximal (minimal in absolute value).

Such a criterion for the step by step determination of the reaction path is not, however, entirely satisfactory. The reason for this consists in the local character of the criterion that is not sufficient to ensure that the reaction path starting at the intial reactant will indeed end at the desired product. This suggests that in proposing a satisfactory formulation of the least motion principle, the above local criterion of minimal changes in electron configuration has to be replaced by the global one. The simplest way to do this is to introduce a curve integral summing up the contributions of the local second derivative along a chosen path in the direction tangential to this path (Eq. 28)

$$L = \int_R^P s\nabla\nabla Ks dl \tag{28}$$

Taking into account that the integral (28) represents a functional, whose value depends on the actual form of the path along which we integrate, the "best" reaction path satisfying the requirement of minimal (global) changes in electron configuration can be naturally determined from the variational condition (29), which can be thus regarded as the desired general formulation of the least motion principle.

$$\delta L = 0 \qquad (29)$$

To solve this problem we proposed a simple numerical method [91] on the basis of which the model was applied to several selected pericyclic reactions. In order to maintain close correspondence with the example already presented of the butadiene to cyclobutene cyclisation, the above formalism will first be demonstrated only for this specific pericyclic process. On the basis of the idealised reaction scheme (Scheme II), the whole process is quite straightforward and consists in converting corresponding structural formulae into the approximate wave functions. These functions are then, in the next step, converted into the topological density matrix $\rho(\vartheta, \varphi|x_1)$ and the similarity index K. On the basis of these indices it is then possible to solve the variational problem (29), and, after transforming the found reaction path into the usual More O'Ferrall variables Q_1, Q_2, the resulting trajectory can be drawn into the More O'Ferrall diagrams. In our case the results are given in Figs. 6 and 7.

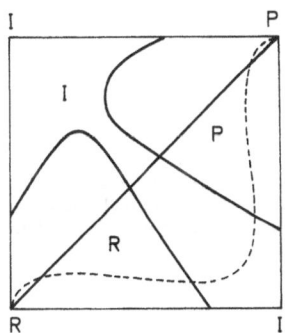

Fig. 6. The partitioning of a modified More O'Ferrall diagram with the corresponding reaction path for thermally forbidden disrotatory cyclisation of butadiene to cyclobutene. The extent of electron reorganisation, measured by the value of the functional L, is -0.48 along this reaction path

Let us discuss now the most important conclusions that can be deduced from these figures. First, the most important conclusion concerns the comparison of the values of functional L along the optimal allowed and forbidden reaction paths. As can be seen, the value for the allowed conrotatory cyclisation is lower in absolute value than in the forbidden one. This confirms the intuitive expectation of the least motion principle that the extent of electron reorganisation should be smaller in allowed reactions than in the forbidden ones. On the basis of this primary test of reliability of the proposed model it is, in the next step, possible to start with the analysis and the classification of the reaction mechanisms for both individual reactions. Especially interesting in this connection is again the thermally forbidden disrotatory cyclisation. The reason for this lies, as already shown above, in the specific form of the partitioning of the More O'Ferrall diagrams which implies that no reaction path can transverse from the region of the reactant into the region of the product without avoiding the region of the intermediate. The reaction mechanism is therefore classified as non-concerted. What is, however, especially interesting is that this intrinsic non-concertedness is accompanied also by considerable assynchronicity of the reaction path which suggests the stepwise nature of the process. This result is quite encouraging since the above picture of the reaction mechanism fits well not only with the qualitative expectations of the Woodward–Hoffmann rules but also with the results of semiempirical MINDO calculations by Dewar [74] and, indirectly, also with the results of more sophisticated newer studies [75, 76]. In these studies the authors report that a symmetrical critical structure corresponding to a forbidden disrotatory reaction is not the true saddle point but the second order saddle point. This implies, however, [92] that there is another assynchronous reaction path of lower energy which could thus correspond to what we found.

If now the same analysis is applied to allowed conrotatory reaction then the corresponding More O'Ferrall diagram (Fig. 7) is in this case compatible with both concerted and nonconcerted mechanisms and the discrimination between them requires a knowledge of the corresponding reaction path.

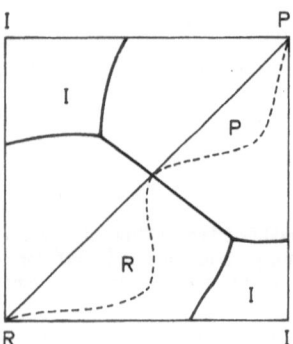

Fig. 7. The partitioning of a modified More O'Ferrall diagram with the corresponding reaction path for thermally allowed conrotatory cyclisation of butadiene to cyclobutene. The extent of electron reorganisation, measured by the value of the functional L, is -0.28 along this reaction path

As can be seen from the figure, the actual reaction path confirms the qualitative prediction of the Woodward–Hoffmann rules and the reaction is to be classified as concerted even if the path itself is not ideally synchronous. It is, however, interesting that despite the assynchronicity of the path the critical structure corresponding to the transition of the system from the region of the intermediate into the region of the product is quite close to the ideally symmetrical structure expected and also found by various quantum chemical methods.

In a similar way, it would be possible to analyse the mechanisms of any other pericyclic reaction, provided the structure of the intermediate is known with sufficient certainty. This requirement can probably be satisfied for the reactions of parent unsubstituted systems where, owing to low polarity of the system, the intermediate can be reasonably approximated by the biradical structures. A slightly more complex situation can occur, however, in the case of substituted skeletons, where the substituent-induced polarity may increase the contributions of zwitterionic structures. The proposed model is, nevertheless, formulated quite generally so that it can be applied to any biradicaloid intermediate described in the sense of the Salem and Rowland theory [93] by the wave function in the form of a linear combination of two zwitterionic and one biradical structure (Eq. 30):

$$\Psi_I = a\Psi_{Z_1} + b\Psi_{Z_2} + c\Psi_B \tag{30}$$

This opens up the possibility of a systematic investigation of pericyclic reactions not only for model cases of parent unsubstituted systems, but for inclusion if zwitterionic contributions also enable the analysis of the eventual mechanistic changes induced by the polar substitution. As an example, the push-pull substituted Diels-Alder system will be analysed, in which the diene component is substituted in position 1 by a donor, and dienophilic component in position 6 by an acceptor substitution. In order to avoid the problems with the relative wieght of individual limiting structures of the intermediate (Eq. 30), the coulombic integrals modelling the substitution in the HMO wave function were arbitrarily set to $\alpha = 3\beta$ and $\alpha = -3\beta$ so that there is sufficient polarity in the system to secure the approximation of the intermediate by pure zwitterionic structure Z_1.

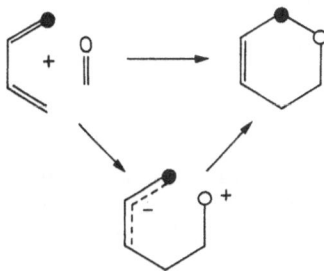

Scheme 4.

The resulting More O'Ferrall diagram corresponding to the allowed $s + s$ mechanism is given in Fig. 8. The most important feature of this diagram is that after a nearly synchronous approach at the initial stages of the reaction, the reaction path becomes considerably asynchronous and hits the region of the intermediate so that the mechanism should be classified as nonconcerted.

This result is again very interesting since even if the mechanism of the parent Diels-Alder reaction is still the focus of the discussion, the majority of both theoretical [94, 95] and experimental results [96, 97] for substituted systems are rather consistent with a stepwise nature of the process which just corresponds to what we found.

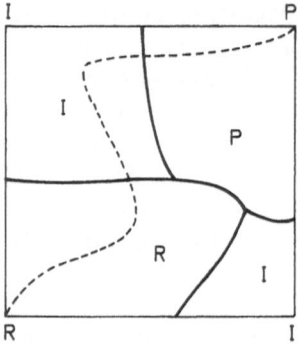

Fig. 8. The partitioning of a modified More O'Ferrall diagram with the corresponding reaction path for the allowed $s + s$ Diels-Alder cyclisation of 1-substituted butadiene with substituted ethene. The extent of electron reorganisation, measured by the value of the functional L, is $- 0.34$ along this path

3 Summary

Although a number of examples presented above convincingly demonstrates that the applications of similarity ideas in the theory of pericyclic reactions are very broad and fruitful, the possibilities of the approach are still far from being exhausted. The whole formalism can be, namely, quite straightforwardly extended beyond the scope of simple processes involving one intermediate to more complex reaction schemes. This would be especially interesting for catalytic reactions where such a generalisation would open the way to the systematic study of the mechanisms of such reactions and to the subsequent prediction of structural effects responsible for the optimal catalytic activiity in a given process. Another interesting application of the proposed approach concerns, for example, the generalisation of the original HMO-based similarity index to the level of more sophisticated semiempirical or even ab initio techniques. This would open the way to more realistic estimates of the similarity indices required for extension of the similarity approach beyond the scope of simple model calculations to the analysis of real chemical systems. The research along this line is being pursued in our laboratory and the results will be published elsewhere.

Acknowledgements. The author wishes to thank the American Chemical Society and J.C. Balzer AS for kind permission to reproduce the material from his own studies published in J. Chem. Inf. Comp. Sci. and J. Math. Chem.

References

1. Rouvray DH (1995) Chapter 1, Topics in Current Chemistry 173: 1–30
2. Rouvray DH (1990) In: Johnson MA, Maggiora G (eds) Concepts and applications of molecular similarity. J Wiley, New York, p 15; Rouvray DH (1992) J Chem Inf Comp Sci 32: 580

3. Rouvray DH (1993) The necessity of analogies in the development of Science, plenary lecture at 1st Girona seminar on molecular similarity, July 8–10, Girona Spain
4. Mezey PG (1991) J Math Chem 7: 39
5. Mezey PG (1992) J Chem Inf Comp Sci 32: 650
6. Mezey PG (1992) J Math Chem 11: 27
7. Mezey PG (1993) In: Shape in Chemistry. An introduction to molecular shape and topology, VCH Publishers, New York
8. Amoros C, McWeeny R (1991) Theochem 227: 1
9. Hopfinger JA, Bierke B (1990) In: Johnson MA, Maggiora G (eds) Concepts and applications of molecular similarity. J. Wiley, New York, p 173
10. Dugundji J, Ugi I (1973) Top Curr Chem 39: 19
11. Ugi I, Wochner MA, Fontain E, Bauer J, Gruber B, Karl R (1990) In: Johnson MA, Maggiora G (eds) Concepts and application of molecular similarity. J. Wiley, New York, p 239
12. Jochum C, Gasteiger J, Ugi I, Dugundji J (1982) Z Naturforsch 376: 1205
13. Bauer J, Herges R, Fontain E, Ugi I (1985) Chimia 39: 43
14. Lawson AJ (1992) J Chem Inf Comp Sci 32: 675
15. Gasteiger J, Ihlenfeld WD, Fick R, Rose JR (1992) J Chem Inf Comp Sci 32: 700
16. Judson PN (1992) J Chem Inf Comp Sci 32: 657
17. Kier LB (1987) In: Hadzi D, Jerman-Blazic B (eds) Drug design and Toxicology. Elsevier, Amsterdam, p 13
18. Richards WG (1993) Pure & Appl Chem 65: 231
19. Cooper DL, Mort KA, Allan N, Kinchington D, McGuigan CH (1994) J Am Chem Soc (in press)
20. Good AC (1992) J Mol Graphics 10: 144
21. Carbo R, Calabuig B (1992) Theochem 254: 517
22. Sanz F (1993) An approach to explore electrostatic similarity between biomolecules, plenary lecture at 1st Girona seminar on molecular similarity, July 8–10, Girona Spain
23. Carbo R, Calabuig B (1992) Int J Quant Chem 42: 1695
24. Carbo R, Calabuig B (1992) J Chem Inf Comp Sci 32: 600
25. Mehlhorn A, Fratev F, Polansky OE, Monev V (1984) Match 15: 3
26. Fratev F, Polansky OE, Mehlhorn A, Monev V (1979) J Mol Struct 56: 245
27. Polansky OE, Derflinger G (1967) Int J Quant Chem 1: 379
28. Carbo R, Leyda L, Arnau M (1980) Int J Quant Chem 17: 1185
29. Carbo R, Calabuig B (1989) Comp Phys Commun 55: 117
30. Bowen-Jenkins PE, Cooper DL, Richards WG (1985) J Phys Chem 89: 2195
31. Hodgkin EE, Richards WG (1986) J Chem Soc Chem Commun 1342
32. Hodgkin EE, Richards WG (1987) Int J Quant Chem Quant Biol Symp 14: 105
33. Ponec R (1987) Coll Czech Chem Commun 52: 555
34. Ponec R, Strnad M (1993) Croat Chim Acta 66: 123
35. Cooper DL, Allan NL (1989) J Comp Aided Mol Design 3: 253
36. Cooper DL, Allan NL (1992) J Am Chem Soc 114: 4773
37. Carbo R, Calabuig B (1992) Int J Quant Chem 42: 1681
38. Cioslowski J, Fleischmann ED (1991) J Am Chem Soc 113: 64
39. Ortiz VJ, Cioslowski J (1991) Chem Phys Lett 185: 270
40. Cioslowski J (1991) J Am Chem Soc 113: 6756
41. Cioslowski J (1993) Atomic contributions to molecular similarity. Plenary lecture at 1st Girona seminar on molecular similarity, July 8–10, Girona, Spain
42. Clar E (1972) The aromatic sextet. J. Wiley, London
43. Ponec R (1984) Coll Czech Chem Commun 49: 455
44. Woodward RB, Hoffmann R (1970) The conservation of orbital symmetry. Academic Press, New York
45. Rice FO, Teller E (1938) J Chem Phys 6: 489
46. Ponec R, Strnad M (1990) Coll Czech Chem Commun 55: 896
47. Ponec R, Strnad M (1990) Coll Czech Chem Commun 55: 2583
48. Ponec R (1987) Z Phys Chem (Leipzig) 268: 1180
49. Ponec R (1989) Z Phys Chem (Leipzig) 270: 365
50. Ponec R, Strnad M (1990) Coll Czech Chem Commun 55: 622
51. Ponec R, Strnad M (1991) J Phys Org Chem 4: 701
52. Ponec R, Strnad M (1992) J Phys Org Chem 5: 764
53. Ponec R, Strnad M (1990) Coll Czech Chem Commun 55: 2363
54. Ponec R, Strnad M (1992) Int J Quant Chem 42: 501

55. Ponec R, Strnad M (1993) Coll Czech Chem Commun 58: 1751
56. Ponec R, Strnad M (1992) J Chem Inf Comp Sci 32: 693
57. Ponec R, Strnad M (1991) J Math Chem 8: 108
58. Ponec R (1993) J Chem Inf Comp Sci 33: 805
59. Ponec R, Strnad M (1994) Coll Czech Chem Commun (in press)
60. Michl J (1970) In: Eyring H, Henderson D, Jost W (eds) Physical Chemistry. An advanced Treatise Vol. 7. Academic Press, New York
61. Huisgen R, Sturm HJ, Wagenhofer H (1962) Z Naturforsch 176: 202
62. Berson J (1980) Acc Chem Res 5: 406
63. Baldwin JE, Andrist HA, Pinschmidt RK Jr (1972) Acc Chem Res 5: 402
64. Gajewski JJ (1980) Acc Chem Res 13: 142
65. Trost B, Miller ML (1988) J Am Chem Soc 110: 6378
66. von Doering WE, Roth WR, Brinckmann R, Figge L, Lennarz H, Fessner WD, Prinzbach H (1980) Chem Ber 121: 1
67. van Mele B, Huybrechts G (1987) Int J Chem Kinet 19: 363
68. Lowe JP (1974) J Chem Educ 51: 785
69. More O'Ferrall RA (1970) J Chem Soc (B) 274
70. Salem L (1982) In: Electrons in Chemical Reactions. J. Wiley, New York, Chapter 2
71. Mezey PG (1981) Theor Chim Acta 58: 309
72. Ponec R (1985) Coll Czech Chem Commun 50: 1121
73. Tal Y, Bader RFW, Nguyen-Dang TT, Ojha M, Anderson SG (1981) J Chem Phys 74: 5162
74. Dewar MJS, Kirschner S (1974) J Am Chem Soc 96: 5244
75. Bofill JM, Gomez J, Olivella S (1988) Theochem 163: 285
76. Breulet J, Schaefer III HF (1984) J Am Chem Soc 106: 1221
77. Houk KN, Li Yi Evanseck JD (1992) Angew Chem Int Ed 31: 682
78. Dewar MJS (1984) J Am Chem Soc 106: 7892
79. Segal G (1979) J Am Chem Soc 96: 7892
80. Bernardi F, Botoni A, Robb MA, Schlegel HB, Tonachini G (1985) J Am Chem Soc 107: 2260
81. Dewar MJS, Olivella S, Rzepa H (1978) J Am Chem Soc 100: 5650
82. Dewar MJS, Olivella S, Stewart JP (1986) J Am Chem Soc 108: 5771
83. Woodward RB, Katz TJ (1959) Tetrahedron 5: 70
84. Houk KN (1989) Pure & Appl Chem 61: 643
85. Borden WT, Loncharich RJ, Houk KN (1984) Ann Rev Phys Chem 39: 213
86. Loncharich RJ, Brown FK, Houk KN (1989) J Org Chem 54: 1129
87. Townshend RE, Rammuni G, Segal G, Hehre WJ, Salem L (1976) J Am Chem Soc 98: 2190
88. von Doering WE, Neumann NM, Hasselmann D, Kaye RL (1972) J Am Chem Soc 94: 3833
89. Li Yi Houk KN (1993) J Am Chem Soc 115: 7478
90. Banthorpe DV (1963) In: Hughes ED (ed) Reaction mechanisms in Organic Chemistry, Vol 2, Elimination reactions. Elsevier, Amsterdam
91. Strnad M (1990) PhD dissertation Institute of Chemical Process Fundamentals, Czechoslovak Academy of Sciences, Prague
92. Murrel JN, Laidler KH (1968) Trans Faraday Soc 64: 371
93. Salem L, Rowland C (1972) Angew Chem Int Ed 11: 92
94. Choi YJ, Lee I (1989) J Chem Soc Faraday Trans 2 85: 867
95. Dewar MJS (1989) Theochem 200: 301
96. Dewar MJS, Pierini A (1984) J Am Chem Soc 106: 203
97. Huybrechts G, Poppelsdorf H, Maesschalck L, van Mele B (1984) Int J Chem Kinet 16: 93

Mapping of Molecular Electric Potentials and Fields

P.C. Mishra* and Anil Kumar

Department of Physics, Banaras Hindu University, Varanasi 221 005, India

Table of Contents

1 Introduction 28

2 Method of Computation 29

3 Applications 30
   3.1 Hydrogen Bonding vs Electric Field 30
   3.2 Study of Molecular Anions and Cations 36
   3.3 Study of Drugs 37
      3.3.1 Cardiotonic Drugs 37
      3.3.2 Anti-Viral Drugs 39

4 Conclusion 41

5 References 42

An overview of earlier work on electric potentials and fields with emphasis on the latter, where these quantities have been used to compare molecular binding properties, is presented. Electric field maps of different types of molecules were calculated using optimized geometries, Mulliken charges and the probe point dipole method. The computed electric fields and experimentally observed hydrogen bonding parameters (α and β) are found to be correlated. It shows that electric field can be used as a reliable descriptor of hydrogen bonding ability of atoms in molecules. The insight obtained by application of electric field mapping to different types of drugs as to their mechanisms of action is discussed.

---

* Author to whom all correspondence should be addressed

Topics in Current Chemistry, Vol. 174
© Springer-Verlag Berlin Heidelberg 1995

1 Introduction

Molecular similarity is a valuable concept in drug design [1]. Thus in view of the lock and key model of substrate-receptor steric complimentarity, a drug and the substrate, or a known lead drug and others under development, should have appropriate similarity. Three dimensional geometry and similarity indices defined in terms of electron density distributions have been used for comparing molecules quite successfully [1]. Similarity and differences between molecules need also be evaluated at the level of their interactions with the appropriate receptor. The following two types of intermolecular interactions are usually dominant in ligand-receptor binding [2]: (i) electrostatic e.g. hydrogen bonding and (ii) hydrophobic, introduced by the aqueous chemical environment. Of these two types of interactions, the former is usually of greater importance. Quantitatively accurate parameters which can be used to characterize molecular electrostatic similarity are of great importance since they can be correlated with the binding properties of the molecules in question. Molecular electric potential (MEP) mapping has been widely used quite successfully to explain electrostatic interactions of a variety of molecules [3–12]. Molecular similarity between compounds which interact with the same receptor can be understood in terms of their MEP maps [13–16]. This point has been demonstrated by Sanz et al. [16] by examining MEP maps of several indole and tryptamine derivatives. Pèpe et al. [17] have studied the utility of MEP in drug design. Arteca and Mezey [18] have proposed several possible graph theoretical approaches to rationalize similarities between molecular surfaces including those of MEP. Molecular electric fields (MEF) have rarely been studied though in several papers their significance from the point of view of intermolecular interactions and recognition has been emphasized [1, 2, 19–21]. MEF for different forms of DNA were calculated by Laveri and Pullman [22] and Pullman et al. [23]. They compared the features shown by MEP and MEF maps and found that, while the deepest MEP regions lie in the grooves of DNA, the strongest MEF regions are associated with the phosphate groups [22, 23]. They correlated this MEF feature with observed preferential hydration of phosphate groups of DNA [22, 23]. Electric field patterns represented by their directions at different points around the molecules were obtained for water and formaldehyde by Peinel et al. [24] Náray-Szabó assumed transferability of MEF values associated with different types of bonds and lone pair contributing atoms [2, 25]. He also correlated MEF with hydrophobicity [2]. Regression analyses were performed by Náray-Szabó for many molecules, showing even somewhat better structure-activity correlations than those obtained by the Hansch method, besides the advantage of conceptual clarity at the level of intermolecular interactions [2, 25].

The success achieved in the above mentioned studies argues strongly in favour of more and better work on MEF mapping. As MEF is a vector quantity, its presentation around molecules poses some problem. In some of the earlier work, MEF directions and dipole-molecule interaction energies were given

[24, 26–30] while in certain other studies, only MEF magnitudes were presented [22, 23]. Colour graphics has been used in a study of MEF map of uracil [31]. It is desirable to study the utility of MEF as a parameter to explain molecular activities, e.g. those of drugs. The present article was mainly intended to meet the following two objectives: (i) to examine if MEF can be used as a reliable index to compare molecules, particularly in relation to hydrogen bonding, and (ii) to overview the earlier work on MEF where calculations were performed at a certain level consistently for several types of molecules. This study is necessary before MEF is used as a standard variable in drug design and other applications.

2 Method of Computation

Molecular electrostatic potential (MEP) $V(\mathbf{r})$ at a point $\mathbf{r}$ is defined as

$$V(\mathbf{r}) = \sum_i Z_i / |\mathbf{R}_i - \mathbf{r}| - \int \rho(\mathbf{r}') \, d\mathbf{r}' / |\mathbf{r}' - \mathbf{r}|$$

where Z_i is the charge on nucleus i located at $\mathbf{R}_i$ and $\rho(\mathbf{r})$ is the electronic charge density function of the given molecule. When the molecular charge distribution is represented in terms of partial point charges q_j located at the atomic sites j, MEP can be obtained using the expression

$$V(\mathbf{r}) = \sum_j q_j / |\mathbf{r}_j - \mathbf{r}|.$$

Electric field can then be computed using

$$E(\mathbf{r}) = -\nabla V(\mathbf{r}).$$

Despite the fact that MEP and MEF are related by the above equation, their spatial distributions may be quite different, and sometimes MEF maps may be more useful than MEP maps and vice versa. It is suggested that preferred binding sites of cations to molecules may be brought out reliably by MEP maps while, for the study of interactions of molecules with neutral dipolar species, MEF maps may be particularly useful [2, 23]. Due to the vectorial character of MEF, its magnitude and direction can both be employed to evaluate molecular similarity more rigorously than what is possible with MEP. If E_A and E_B are the MEF of two molecules A and B at the corresponding point, one may use their scalar or vector product to evaluate similarity between the molecules [1].

We have used the following method for calculating electric fields. Suppose the magnitude of electric field due to the charge distribution of a molecule at a point in its vicinity is E. If a point dipole having moment p is placed at that point, its potential energy of interaction with the electric field would be given by

$$W = -pE \cos \theta$$

where θ is the angle between the direction of the point dipole and that of the electric field. If the dipole is allowed to rotate freely, it would orient itself along the minimum energy direction ($\theta = 0$) which would also be the direction of the field. Then

$$E = -W/p.$$

In some of our earlier studies, we presented values of W [26–30]. MEP maps are usually obtained in terms of contours of equal potential. However, maps giving MEP values on van der Waals surfaces have been found to be more useful for structure-activity correlation [2]. For the same reason, it is preferable to have MEF maps giving values on van der Waals surfaces. Accessibility of atomic sites is an important aspect in connection with MEP studies [32]. When MEP or MEF maps are obtained giving values on van der Waals surfaces, accessibility of sites is easily accounted for. We compute MEF at points which are at distances $(R_i + R_j + \Delta)$ from the atomic sites where R_i and R_j are van der Waals radii of atoms of the molecule under study and those of the charged ends of the point dipole (taken to be 1 Å each) respectively, and Δ is a parameter which can be fixed at suitable values in order to generate desired surfaces of closest distance of approach (CDA) between the dipole and atoms of the molecule.

3 Applications

3.1 Hydrogen Bonding vs Electric Field

Hydrogen bonding is considered to be mainly controlled by electrostatic interactions. According to a fairly successful empirical formula due to Allen [33], interaction energy due to the (AH . . . B) hydrogen bond between two sites A and B is proportional to the AH bond moment. Hydrogen bonding abilities of atoms of a molecule, e.g. those of a solvent, are measured by two parameters [34–37]: (i) hydrogen bond donating parameter (α), and (ii) hydrogen bond accepting parameter (β). Utility of calculated minimum electrostatic potentials V_{min} has been investigated by Murray et al. [37] in relation to the hydrogen bond accepting parameters (β) in several cases. Study of nucleophilic processes by MEP mapping is not quite straightforward. However, Murray and Politzer [35] have shown that positive maximum MEP values (V_{max}) computed at molecular surfaces defined by 0.002 electron bohr^{-3} contour of electron density, which enclose 95% of total electronic charges, can be satisfactorily correlated with the parameter (α). In another study, Murray et al. [38] have shown that the best correlation between α and V_{max} is obtained if the molecular surface is defined by 0.001 electron bohr^{-3} contour of electron density which lies at a distance close to the van der Waals radius of hydrogen.

In order to examine if MEF values can describe hydrogen bonding abilities of atoms represented by the parameter (α), we have calculated MEF maps of the molecules which were studied by Murray and Politzer [35]. We optimized geometries of the molecules using the MNDO method [39], and charge distributions were obtained using the Mulliken population analysis. Our calculated electric fields in front of the OH bonds of eleven molecules along with the measured values of the parameter (α) are presented in Table 1. The MEP (V_{max}) values obtained by Murray and Politzer [35] and scaled values of V_{max} so as to make the present highest MEF value and their highest MEP value numerically equal, for convenience of graphical presentation, are also shown in Table 1. MEF maps of some of the molecules included in this study are shown in Figs. 1 and 2. The points on CDA surfaces in front of the OH bonds, MEF values at which are included in Table 1, are each marked by an asterisk (*). Although

Table 1. Electric fields (F), electrostatic potentials (V_{max}) and hydrogen bond donating parameters (α) for molecules with hydroxyl groups as hydrogen bond donors

Sl. no.	Molecule	F (Volt/nm)	V_{max} (kcal/mol) unscaled (scaled)[a]	α^b
1	$(CF_3)_2$ CHOH	6.78	49.6 (6.78)	1.96
2	CF_3CH_2OH	4.32	48.6 (6.64)	1.51
3	H_2O	2.48	38.0 (5.19)	1.17
4	CH_3COOH	3.02	39.0 (5.33)	1.12
5	CH_3OH	2.41	37.3 (5.10)	0.93
6	$HO(CH_2)_2OH$	2.44	37.1 (5.07)	0.90
7	CH_3CH_2OH	2.27	35.6 (4.87)	0.83
8	$CH_3(CH_2)_3OH$	2.34	34.6 (4.73)	0.79
9	$CH_3(CH_2)_2OH$	2.06	35.6 (4.87)	0.78
10	$(CH_3)_2$ CHOH	2.07	35.0 (4.78)	0.76
11	$(CH_3)_3COH$	1.97	31.4 (4.29)	0.68

[a] From [35]. Scaling was made such that the highest values of V_{max} and F are equal in magnitude (Fig. 3)
[b] From [34]

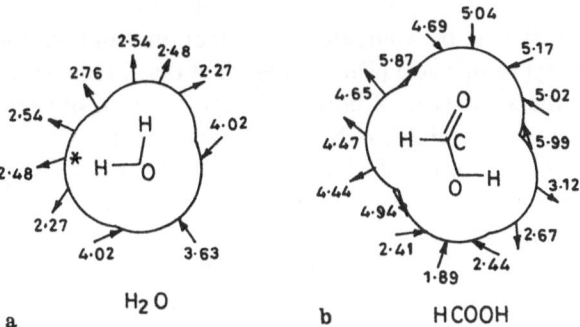

Fig. 1a, b. Electric field map (Volt/nm) of: **a** water; **b** formic acid

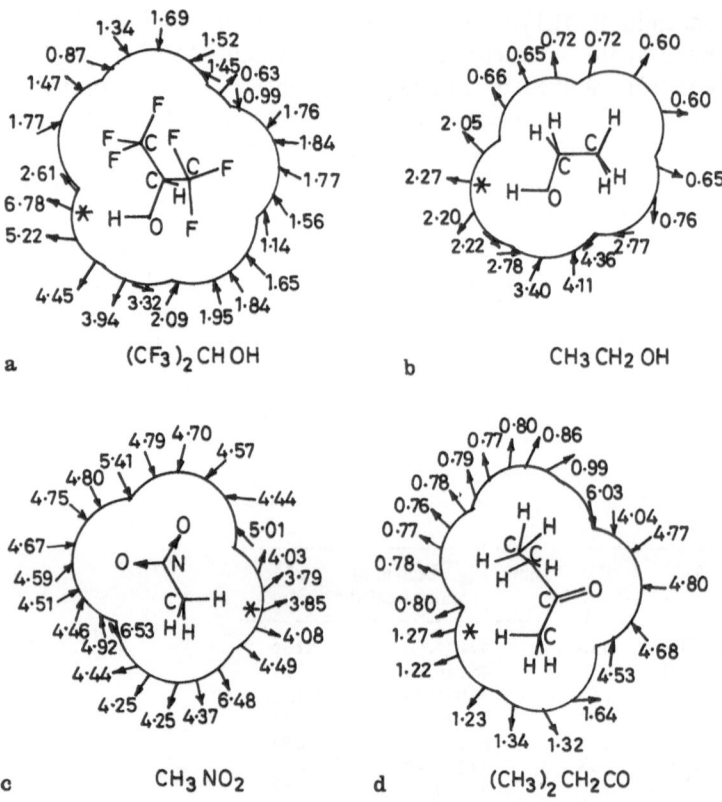

Fig. 2a–d. Electric field maps (Volt/nm) of hydrogen bond donating molecules: **a** $(CF_3)_2$ CHOH; **b** CH_3CH_2OH; **c** CH_3NO_2; **d** $(CH_3)_2CH_2CO$

interaction energies of the molecule of water with a dipole and its electric field patterns (directions) are available in the literature [24, 27], its MEF maps giving both magnitudes and directions of field are not available and hence we are including the MEF map of water here (Fig. 1). The MEF value in front of the OH bond of formic acid (Fig. 1) can be compared with that in front of the corresponding bond of methyl formic acid (Table 1). We find that MEF can be quite different in front of the same type of bond (e.g. OH bond) in different molecules (Table 1), and thus transferability of MEF values in such cases would not be justified. Variation of the present calculated electric field values and the scaled V_{max} values due to Murray and Politzer [35] with the observed values of the parameter (α) are shown in Fig. 3a.

Electric field values in front of CH bonds of certain molecules for which the hydrogen bonding (α) parameters are available corresponding to situations where alkyl groups are the hydrogen donors, are presented in Table 2, and MEF maps of two of these molecules are shown in Fig. 2. The V_{max} values due to Murray and Politzer (both scaled and unscaled) [35] as well as the observed (α)

Table 2. Electric fields (F), electrostatic potentials (V_{max}) and hydrogen bond donating parameters (α) for molecules with alkyl groups as hydrogen bond donors

Sl. no.	Molecule	F (Volt/nm)	V_{max} (kcal/mol) unscaled (scaled)[a]	α[b]
1	CH_3NO_2	3.85	32.9 (3.85)	0.22
2	CH_3CN	2.07	27.4 (3.21)	0.19
3	$(CH_3)_2CO$	1.34	17.3 (2.02)	0.08
4	$(CH_3)_2CH_2CO$	1.27	18.4 (2.15)	0.06

[a] From [35]. Scaling was made such that the maximum values of V_{max} and F are equal in magnitude (Fig. 3)
[b] From [34]

parameters are also presented in Table 2. Variation of the observed (α) values of these molecules with our calculated MEF values and with the scaled values of V_{max} due to Murray and Politzer [35] is plotted in Fig. 3b. Murray and Politzer have represented the variation of (α) with V_{max} by a straight line [35]. Variation of MEF with (α), however, seems to deviate significantly from straight line behaviour. Our MEF values and the V_{max} values due to Murray and Politzer [35] given in Table 1 and Fig. 3a correlate with the observed (α) parameters equally satisfactorily (linear correlation coefficient 0.96 in each case). But in Table 2 and Fig. 3b, the correlation between V_{max} and observed (α) parameters is somewhat better (linear correlation coefficient 0.98) than that between our MEF and the observed (α) parameters (linear correlation coefficient 0.88).

Correlations have been shown to exist between hydrogen bond accepting parameters (β) and calculated minimum electrostatic potentials V_{min} near electron-rich sites, e.g. nitrogen and oxygen, for several molecules by Murray et al. [37]. In order to examine if MEF values can also be correlated with the (β) parameters, we have computed MEF maps for two types of molecules which

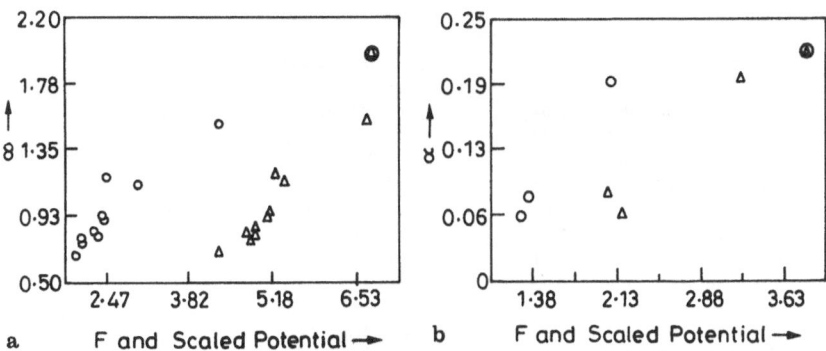

Fig. 3a, b. Plot of hydrogen bond donating parameter (α) with electric field (F) (Volt/nm), shown by o, and scaled potential V_{max} (kcal/mol), shown by Δ: **a** for molecules containing OH group (Table 1); **b** for molecules containing alkyl groups (Table 2)

were studied by Murray et al. [37], namely some nitrogen-containing heterocycles and alkyl ethers. The molecules of one of these groups have a closed ring while those of the other group have an open chain each. Optimized geometries and Mulliken charges obtained employing the MNDO method were used in these MEF calculations. The MEF map of one molecule of each of these two classes is presented in Fig. 4. Our calculated MEF values, the V_{min} values due to Murray et al. [37] along with the scaled magnitudes of V_{min} and the observed (β) parameters for these two groups of molecules are presented in Tables 3 and 4. As in the case of (α) parameters, the scaling of $|V_{min}|$ was done for each of these two groups of molecules separately so as to make the respective highest MEF and the highest $|V_{min}|$ values numerically equal, for the sake of convenience of graphical

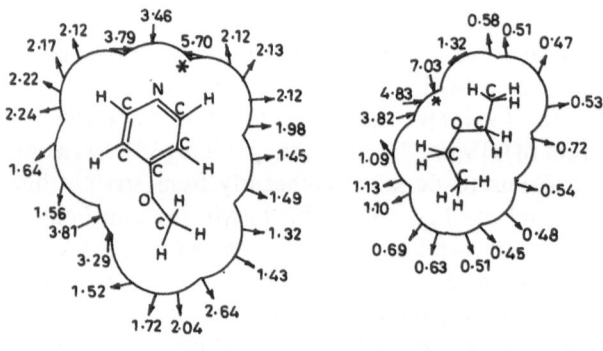

a   4-Methoxy Pyridine   b $H_3CH_2COCH_2CH_3$

Fig. 4a, b. Electric field maps (Volt/nm) of hydrogen bond accepting molecules: **a** 4-methoxy pyridine; **b** $H_3CH_2COCH_2CH_3$

Table 3. Electric fields (F), electrostatic potentials (V_{min}) and hydrogen bond accepting parameters (β) for some nitrogen-containing heterocycles

Sl. no.	Molecule	F (Volt/nm)[a]	V_{min} (kcal/mol) unscaled (scaled)[b]	β[c]
1	4-(Dimethylamino) pyridine	5.97	-97.3 (6.29)	0.87
2	N-methylimidazole	6.61	-102.3 (6.61)	0.82
3	2,6-Dimethylpyridine	5.73	-93.5 (6.04)	0.76
4	4-Methoxypyridine	5.70	-93.4 (6.03)	0.72
5	3-Methylpyridine	4.96	-92.6 (5.98)	0.68
6	4-Methylpyridine	5.61	-93.1 (6.01)	0.67
7	Pyridine	5.62	-91.0 (5.88)	0.64
8	Pyrimidine	5.41	-82.2 (5.31)	0.48
9	3,5-Dichloropyridine	4.43	-72.1 (4.66)	0.42

[a] Highest MEF value near the nitrogen atom of the ring (Fig. 4a)
[b] Deepest MEP value near the nitrogen atom, from [37]. The scaling was made such that the largest magnitudes of V_{min} and F are equal. In scaling, $|MEP|$ was considered (Fig. 5)
[c] From [34]

Table 4. Electric fields (F), electrostatic potentials (V_{min}) and hydrogen bond accepting parameters (β) for some alkyl ethers

Sl. no.	Molecule	F (Volt/nm)[a]	V_{min} (kcal/mol) unscaled (scaled)[b]	β[c]
1	$(CH_3)_2\,HCOCH\,(CH_3)_2$	7.23	-70.7 (7.23)	0.49
2	$H_3CH_2COCH_2CH_3$	7.03	-65.1 (6.66)	0.47
3	$H_3C(H_2C)_2O(CH_2)_2CH_3$	5.50	-65.7 (6.72)	0.46
4	$H_3CO(CH_2)_2OCH_3$	5.00	-60.3 (6.17)	0.41

[a] Highest MEF value near the oxygen atom (Fig. 4b)
[b] Deepest MEP value near the oxygen atom, from [37]. The scaling was made such that the largest magnitudes of V_{min} and F are equal. In scaling, |MEP| was considered (Fig. 5)
[c] From [34]

presentation (Fig. 5a, b). We find that in both the groups of molecules, our MEF values, the V_{min} values due to Murray et al. [37] and the observed (β) parameters are nicely correlated (Tables 3,4 and Fig. 5a,b). In Table 3 and Fig. 5a, the correlation between the V_{min} values due to Murray et al. [37] and the observed (β) parameters is better (correlation coefficient 0.94) than that between the corresponding values of MEF and (β) (correlation coefficient 0.79 including 3-methylpyridine, 0.85 excluding it). In Table 4 and Fig. 5b, the correlation of the observed (β) parameters and the V_{min} values due to Murray et al. [37] is also somewhat better (correlation coefficient 0.95) than that between our MEF and the (β) values (correlation coefficient 0.88). As we have used only semiempirical partial charges located at the atomic sites in the MEF calculations, while the MEP calculations are based on ab initio wavefunctions, the somewhat inferior correlation obtained between the MEF and observed (β) values is understandable.

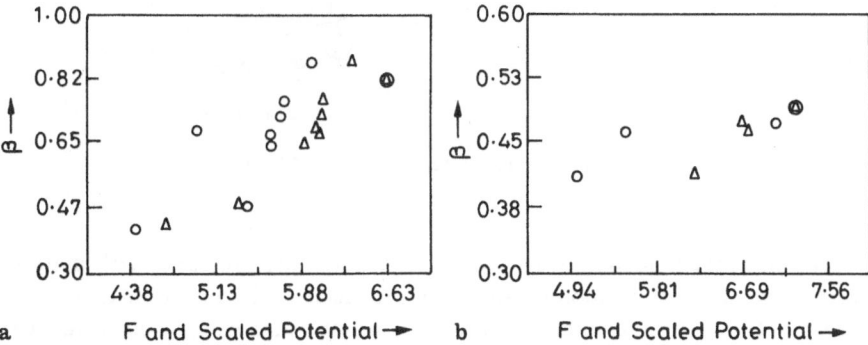

Fig. 5a, b. Plot of hydrogen bond accepting parameter (β) with electric field (F) (Volt/nm), shown by o, and scaled potential |V_{min}| (kcal/mol), shown by Δ: **a** for nitrogen-containing heterocycles; **b** for alkyl ethers

3.2 Study of Molecular Anions and Cations

Anions and cations of molecules play important roles in biology and chemistry. For example, neurotransmitter molecules carry charge in cationic state [40]. Development of techniques to study anions and cations is, therefore, very important. Anions have been successfully studied using MEP mapping [11] but application of this technique to cations poses difficulty [36]. However, MEF mapping is equally straightforward for neutral and ionic forms of molecules. Anions, cations and neutral forms of adenine, guanine, 2-aminopurine and 6-thioguanine have been studied earlier by MEF mapping [41, 42]. We present MEF maps of two ionic species here, namely the anion of formic acid and the cation of adenine (Fig. 6). Geometries of the ionic species were fully optimized using the MNDO method and the MEF maps were drawn on two CDA surfaces using Mulliken charges. The MEF map of neutral formic acid (Fig. 1) and that of the corresponding anion (Fig. 6) may be compared. The MEF map of neutral adenine reported earlier [41] may also be compared with that of its cation (Fig. 6). It is found that on the first CDA surface (closest to the species under study), MEF values vary appreciably as we move away from certain sites to the neighbouring ones, but such prominent variations in MEF are not observed on the next CDA surface which is 2 Å away from the first one, in each case (Fig. 6).

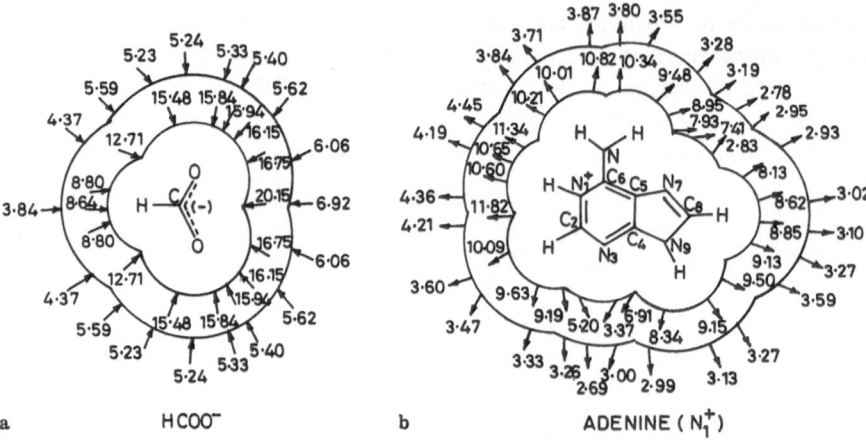

a   HCOO⁻   b   ADENINE (N_1^+)

Fig. 6a, b. Electric field map (Volt/nm) of: **a** formic acid anion; **b** adenine (N_1^+) cation

We make the following observations regarding the features of MEF maps of corresponding neutral, anionic and cationic forms of molecules studied here and earlier [41, 42]. (i) Fields are inward near anions and usually outward near cations. (ii) In neutral molecules, MEF values are high near electron rich sites. This situation parallels the occurrence of deep MEP regions near such sites. (iii) Inward MEF values are further increased near those sites of anions which are associated with high MEF values in the corresponding neutral molecules. When two lone pairs are present in the same region, MEF is maximized in between them as in Fig. 6a. Similar features have been noted earlier in MEP maps of anions [43]. (iv) Regions where MEF are prominently high in neutral molecules are associated with minima of MEF in cations, while the regions near hydrogens, e.g. those of amino group, are associated with high MEF values in cations. This is clear from Fig. 6b, and the results reported earlier [41, 42].

The last observation, (iv) of the above paragraph, is supported by experimental observations on cations, as follows. A hydrogen bond formed between the N_3 position of 8-bromoadenosine and its $5'$-OH group is weakened upon protonation of site N_1 of the molecule [44]. This possibility is supported by the computed field map for the cation of adenine (Fig. 6b). Crystallographic and high resolution NMR studies have revealed valuable information regarding hydrogen bonding between complimentary and mismatch base pairs of DNA. Thus protonation of the N_1 site of adenosine is known to strengthen strongly the hydrogen bond between the N_3 site of cytidine and the N_6 site of adenosine in the cytidine-adenosine mismatch base pair [45–48]. Hydrogen bonds between the O_6 and N_7 sites of neutral guanosine, and the amino group and a proton (at the N_1 site) of a cation of adenosine at low pH, are known to lead to the formation of a stable mismatch base pair which is important for oncogenic sequences [49]. The possibility of formation of such base pairs is strongly supported by the corresponding MEF maps [41, 42] (Fig. 6). The different features exhibited by MEF maps of anions are also supported by experimental data [41, 42]. Thus MEF values can also be reliably used to predict hydrogen bonding properties of anions and cations of molecules.

3.3 Study of Drugs

3.3.1 Cardiotonic Drugs

The enzyme phosphodiesterase (PDE) catalyses conversion of cyclic adenosine monophosphate (cAMP), through hydrolysis of the phosphodiester bond, into AMP [50]. Hyperactivity of the enzyme PDE may cause serious reduction in the level of cAMP which is involved in processes controlling cardiac and smooth muscle contractility, and it may result in congestive heart failure (CHF) [50]. Inhibitors of the enzyme PDE can, therefore, serve as cardiotonics [51]. We have studied two classes of PDE inhibitors: two compounds of the class of bipyridines, namely amrinone and milrinone (Fig. 7) and four compounds of the class of 3-pyridinecarboxylic acids involving different substitutions at the 2-position (Fig. 8). The higher potency of milrinone as compared to that of amrinone is expected to be related to the methyl group substituted at the 2-position [51]. MEF maps of these two molecules computed using crystallographic or MNDO optimized geometries and net charges have revealed the following information [52]. In amrinone, the strongest electric field region lies in the N_1, C_2, C_3, plane

AMRINONE R_1 = H, R_2= NH_2

MILRINONE R_1 = CH_3, R_2 = CN   **Fig. 7.** Molecular structures of amrinone and milrinone

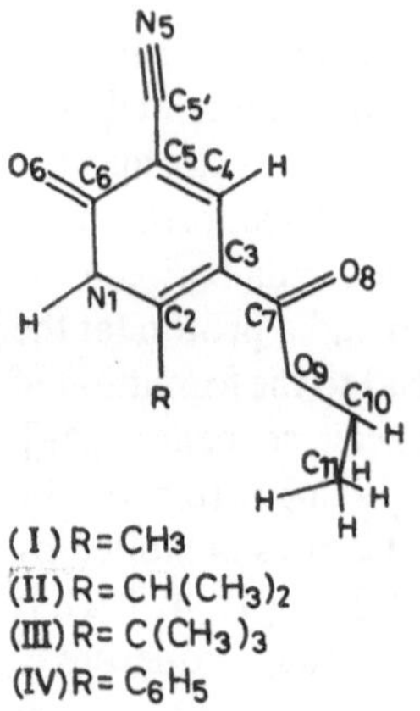

(I) R = CH_3
(II) R = $CH(CH_3)_2$
(III) R = $C(CH_3)_3$
(IV) R = C_6H_5

Fig. 8. Molecular structures of 2-substituted 3-pyridinecarboxylic acids. Different groups (R) are substituted at the 2-position

while, in milrinone, a region with higher fields than those in this plane extends from near the methyl group to around the oxygen atom of the carbonyl group [52]. This difference between their MEF maps may represent differences in the binding patterns of the two drugs with the enzyme.

Substitutions of methyl, isopropyl, *tert*-butyl and phenyl groups at the 2-position in 3-pyridinecarboxylic acid give rise to four drugs with varying potencies, that of the drug with the methyl substitution being the highest [53]. MEF maps of these molecules were computed by us in the same way as those of bipyridines [52]. Average MEF values near the substituents and the oxygen atom of the C_7O_8 bond (Fig. 8) are found to be in the same order as that of the experimentally observed potencies of these drugs [54] (Table 5). The substituents at the 2-position and the oxygen atom of the carbonyl group in each of these molecules, therefore, appear to be involved in binding with the enzyme. In the earlier study [54], significance of electric fields near the substituents at the 2-position only was discussed. The main role of alkyl groups of molecules such as those mentioned above may be to help the drug molecules dock in lipophilic pockets of the receptor. The results for the two classes of drugs discussed above suggest that molecules with methyl substitutions may be able to dock on the receptor surface most effectively [54]. Certain pharmacological models have

Table 5. Electric fields (F) values and cardiotonic activities of 2-substituted 3-pyridinecarboxylic acids

Sl. no.	Molecule[a]	Activity[b]	Range [average][c] of F (Volt/nm)	Range [average][d] of F (Volt/nm)
1	I	Most active	2.44–2.97 [2.75]	3.97–4.58 [4.26]
2	IV	Scarcely active	1.79–1.87 [1.82]	3.87–4.47 [4.17]
3	II	Scarcely active	1.64–1.75 [1.71]	3.59–4.30 [3.93]
4	III	Inactive	1.08–1.20 [1.15]	2.68–3.65 [3.14]

[a] See Fig. 8
[b] From [54]
[c] Electric field strength near the substituent (Fig. 8)
[d] Electric field strength near the oxygen atom of the C_7O_8 bond (Fig. 8)

been proposed regarding the modes of action of PDE inhibitors [55–58]. The Bristol model includes the presence of a carbonyl group in the PDE inhibitor molecule as one of the requirements [58]. Our studies including those mentioned above [52, 54, 59] support this aspect of the Bristol model [58].

3.3.2 Anti-Viral Drugs

Several drugs are obtained by modifying the sugar moiety of guanosine (Fig. 9). In some of these modifications, the sugar ring is replaced by an acyclic group, a well-known drug of the class being acyclovir [60]. In two other drugs of the class also, namely dihydroxybutylguanine (DHBG) and gancyclovir, a similar strategy is adopted (Fig. 9). Acyclovir and gancyclovir are active against the herpes

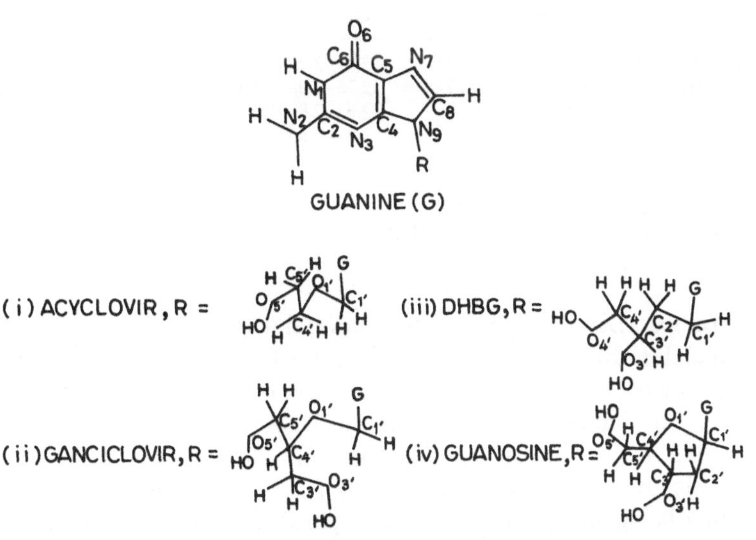

Fig. 9. Molecular structures of acyclic nucleosides and guanosine. Different groups (R) are substituted at the N_9-position of the guanine moiety

simplex virus (HSV) and cytomegalo virus respectively [61, 62]. The activity of DHBG is related to its chirality [63]. The suggested mode of action of these drugs is as follows [64, 65]. In infected cells, the 5'-OH group of these drugs is monophosphorylated by the viral enzyme thymidine kinase (TK), then it is further phosphorylated by the host cell enzyme (GMP kinase) resulting in the formation of a triphosphate which inhibits the viral DNA polymerase [64, 65]. In healthy cells, due to absence of the viral enzyme TK, monophosphorylation and hence further phosphorylation of drugs do not occur. These drugs can, therefore, selectively act on the infected cells [65]. Activity at the 3'-site in the drugs should also be less than that in the natural nucleoside guanosine, so that DNA polymerisation at this site is retarded. These considerations show that activities at the 5'-sites of the drugs should be more than that in guanosine while the corresponding activities at the 3'-sites should be in the reverse order. In acyclovir, there is no 3'-hydroxyl group (instead, there is a dihydro group) and, therefore, reduction of activity of the 3'-site with regard to DNA polymerisation is easily achieved using it [60]. In each of gancyclovir and DHBG, a 3'-OH group is present, and therefore activities at these hydroxyl groups need to be evaluated. MEF values computed by us in front of the O_3,HO_3, and O_5,HO_5, or the corresponding bonds (Table 6) using MNDO geometry optimization and Mulliken charges show that the above mentioned requirements regarding relative activities at the 3' and 5'-sites of the drugs and guanosine are met [66].

Table 6. Electric field (F) values in front of O_3,HO_3, and O_5,HO_5, bonds of acyclic nucleosides and guanosine

Sl. no.	Molecule [a]	Range [average] of F (Volt/nm) O_3,HO_3,	Range [average] of F (Volt/nm) O_5,HO_5,
1	Acyclovir	1.56–1.65 [1.61] [b]	3.49–3.72 [3.59]
2	Ganciclovir	1.14–2.47 [1.81]	2.78–3.08 [2.93]
3	DHBG	2.17–2.48 [2.33]	2.68–2.89 [2.79] [c]
4	Guanosine	3.52–3.70 [3.61]	1.94–2.56 [2.25]

[a] See Fig. 9
[b] Electric field strength near H_4, and H_4,, (Fig. 9)
[c] Electric field strength in front of O_4,HO_4, bond (Fig. 9)

We have studied MEP and MEF maps of several drugs of the class of AZT (azidothymidine) (Fig. 10) which are active against the HIV (Human Immunodeficiency Virus). The relevant biological information in this context, in brief, is as follows. The genetic information stored in the RNA of HIV (retrovirus) is transcribed to DNA, the process being catalysed by the viral enzyme reverse transcriptase which is present only in infected cells [65, 67]. AZT or another drug of this class, after being phosphorylated, is incorporated in the viral DNA [65]. As its 3'-OH group is replaced by the azido group (Fig. 10), viral DNA chain elongation cannot occur at this site [65, 68]. Our calculated MEF maps of

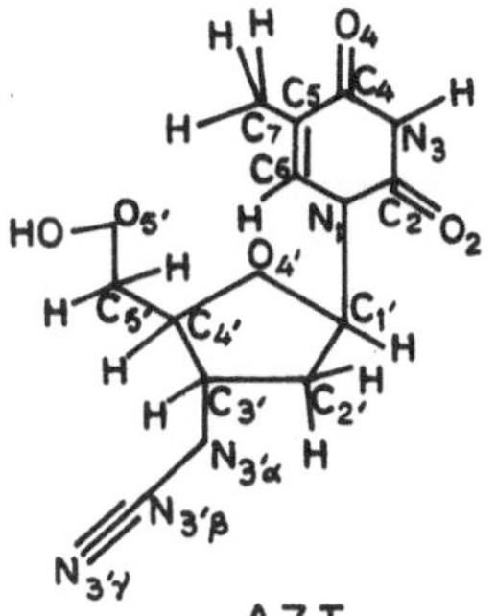

Fig. 10. Molecular structure of azidothymidine (AZT)

several drugs of the class of AZT show that their azido groups would be able to bind weakly with the receptor [69] which catalyses the viral DNA chain elongation [65, 68].

We have also studied MEF maps of several other classes of molecules and drugs, e.g. cardiotonics [59] and other drugs [70], neurotransmitters [71], sweeteners [72] etc. Satisfactory MEF-activity relationships have been found in all these cases.

4 Conclusions

It is found that MEF can be used as a reliable index to compare molecules, particularly in relation to hydrogen bonding. Higher MEF values seem to correspond to stronger hydrogen bonds, and hence higher effects on the relevant processes, and vice versa. Different families of molecules should, however, be treated separately, as found earlier in MEP applications, for meaningful comparison of hydrogen bonding abilities of atoms and structure-activity correlation. Electric fields computed using semiempirical methods (e.g. MNDO) and partial point charges (e.g. Mulliken charges) located at the atomic sites, are found to be satisfactory in this context. This is particularly encouraging from the point of view of drug design and study of large biomolecules where use of accurate ab initio calculations may be prohibitively expensive. However, more accurate charge distribution within the semiempirical molecular orbital framework would be desirable to enhance predictive ability of such calculations.

Acknowledgements. One of the authors (PCM) is thankful to C.S.I.R. (New Delhi) and U.G.C. (New Delhi) for financial support. AK thanks C.S.I.R. (New Delhi) for a Research Associateship.

5 References

1. Dughan L, Burt C, Richards WG (1991) J Mol Struct (Theochem) 235: 481
2. Náray-Szabó G (1989) Int J Quant Chem (Quant Biol Symp) 16: 87
3. Scrocco E, Tomasi J (1973) Top Curr Chem 42: 95
4. Scrocco E, Tomasi J (1978) Adv Quant Chem 11: 115
5. Politzer P, Laurence PR, Jayasuriya K (1985) Environ Health Perspec 61: 191
6. Politzer P, Truhlar DG (eds) (1981) Chemical Applications of Atomic and Molecular Electro-static Potentials. Plenum Press, New York
7. Politzer P, Laurence PR (1984) Carcinogenesis 5: 845
8. Politzer P, Daiker KC, Donnelly RA (1976) Cancer Letters 2: 17
9. Sen KD, Politzer P (1989) J Chem Phys 91: 5123
10. Politzer P, Hedges WL (1982) Int J Quant Chem (Quant Biol Symp) 9: 307
11. Gadre SR, Shrivastava IH (1991) J Chem Phys 94: 4384
12. Gadre SR, Bapat S, Sundararajan K, Shrivastava IH (1990) Chem Phys Lett 175: 307
13. Orozco M, Canela EI, Franco R (1989) Mol Pharmacol 35: 257
14. Richards NGJ, Vinter JG (1991) J Comput Aided Mol Design 5: 1
15. Manaut F, Sanz F, José J, Milesi M (1991) J Comput Aided Mol Design 5: 371
16. Sanz F, Manaut F, Dot T, de Brinas EL (1992) J Mol Struct (Theochem) 256: 287
17. Pèpe G, Siri D, Reboul JP (1992) J Mol Struct (Theochem) 256: 175
18. Arteca GA, Mezey PG (1989) Theoret Chim Acta (Berl) 75: 333
19. Ranghino G, Clementi E (1978) Gazz Chim Ital 108: 157
20. Fruhbeis H, Klein R, Wallmeier H (1987) Angew Chem Int Ed Engl 26: 403
21. Hall G, Smith CM (1988) J Mol Struct (Theochem) 179: 293
22. Laveri R, Pullman B (1982) Nucl Acid Res 10: 4383
23. Pullman A, Pullman B, Laveri R (1983) J Mol Struct (Theochem) 93: 85
24. Peinel G, Frischleder H, Birnstock F (1980) Theoret Chim Acta (Berl) 57: 245
25. Náray-Szabó G (1989) Pharmacochemistry Library 12: 29
26. Mishra PC, Tewari RD (1987) Int J Quant Chem 32: 181
27. Anil Kumar, Mishra PC (1987) Proc Indian Acad Sci (Chem Sci) 99: 113
28. Anil Kumar, Mishra PC (1989) Proc Indian Acad Sci (Chem Sci) 101: 55
29. Anil Kumar, Mishra PC (1990) Int J Quant Chem 38: 11
30. Anil Kumar, Mishra PC (1988) Indian J Biochem Biophys 25: 392
31. Richards NGJ, Price SL (1989) Int J Quant Chem (Quant Biol Symp) 16: 73
32. Pullman A, Pullman B (1981) Quart Revs Biophys 14: 289
33. Allen LC (1975) J Am Chem Soc 97: 6921
34. Kamlet MJ, Abboud JLM, Abraham MH (1983) J Org Chem 48: 2877
35. Murray JS, Politzer P (1991) J Org Chem 56: 6715
36. Sjoberg P, Politzer P (1990) J Phys Chem 94: 3959
37. Murray JS, Ranganathan S, Politzer P (1991) J Org Chem 56: 3734
38. Murray JS, Brinck T, Grice ME, Politzer P (1992) J Mol Struct (Theochem) 256: 29
39. Dewar MJS, Thiel W (1977) J Am Chem Soc 99: 4899
40. Donne-op Den Kelder GM, Haaksma EEJ, Timmerman H (1989) Pharmacochemistry Library 12: 365
41. Mohan CG, Anil Kumar, Mishra PC (1993) Int J Quant Chem 48: 233
42. Mohan CG, Mishra PC (1994) Proc Indian Acad Sci (Chem Sci) 106: 277
43. Gadre SR, Kölmel C, Shrivastava IH (1992) Inorg Chem 31: 2279
44. Sharma R, Lee C, Evans F, Yathindra N, Sundaralingam M (1974) J Am Chem Soc 96: 7337
45. Kalnik MW, Kouchakdjian M, Li BFL, Swann PF, Patel DJ (1988) Biochemistry 27: 100
46. Sowers LC, Fazakerley GV, Kim H, Dalton L, Goodman MF (1986) Biochemistry 25: 3983
47. McConnell B (1974) Biochemistry 13: 4516
48. Hunter WN, Brown T, Anand NN, Kennard O (1986) Nature (London) 320: 552
49. Carbonnaux C, van der Marel GA, van Boom JH, Guschlbauer W, Fazakerley GV (1991) Biochemistry 30: 5449
50. Moos WH, Humblet CC, Sircar I, Rithner C, Weishaar E, Bristol JA, McPhail AT (1987) J Med Chem 30: 1963

51. Robertson DW, Boyd DB (1992) In: Advances in Second Messenger and Phosphoprotein Research, vol. 25, Strada SJ, Hidaka H (Eds) Raven Press, Ltd., New York
52. Anil Kumar, Bhattacharjee AK, Mishra PC (1991) J Mol Struct (Theochem) 251: 359
53. Mosti L, Menozzi G, Schenone P, Dorigo P, Gaion RM, Benetollo F, Bombier G (1989) Eur J Med Chem 24: 517
54. Anil Kumar, Bhattacharjee AK, Mishra PC (1992) Int J Quant Chem 43: 579
55. Erhardt PW, Hagedorn III AA, Sabio M (1988) Mol Pharmacol 33: 1
56. Mohammadi F, Richards NGJ, Guida WC, Liskamp R, Lipton M, Caufield C, Chang G, Hendrickson T, Still WC (1990) J Comput Chem 11: 440
57. Moos WH, Humblet CC, Sircar I, Rithner C, Weishaar RE, Bristol JA, McPhail AT (1987) J Med Chem 30: 1963
58. Bristol JA, Sircar I, Moos WH, Evans DB, Weishaar RE (1984) J Med Chem 27: 1101
59. Mohan CG, Anil Kumar, Mishra PC (1995) J. Mol. Struct (Theochem)
60. Birnbaum GI, Cygler M, Shugar D (1984) Can J Chem 62: 2646
61. Bisacchi GS, Braitman A, Cianci CW, Clark JM, Field AK, Hagen ME, Hockstein DR, Malley MF, Mitt T, Slusarchyk WA, Sundeen JE, Terry BJ, Tuomari AV, Weaver ER, Young MG, Zahler R (1991) J Med Chem 34: 1415
62. Fanhavar P, Nahata MC, Brady MT (1989) J Clin Pharm Ther 14: 329
63. Datema R, Johansson NG, Oberg B (1986) Chem Scr 26: 49
64. Fyfe JA, Keller PM, Furman PA, Miller RL, Elion GB (1978) J Biol Chem 253: 8721
65. Suckling CJ (1991) Sci Progr Edinburgh 75: 323
66. Anil Kumar, Mishra PC (1993) J Mol Struct (Theochem) 288: 151
67. Brown P (1992) Science 54: 1
68. Mohan P (1992) Pharm Res 9: 703
69. Anil Kumar, Mishra PC (1992) J Mol Struct (Theochem) 277: 299
70. Anil C Nair, Mishra PC, J Mol Struct (Theochem) (1994)
71. Santosh C, Mishra PC (1994) J Mol Struct (Theochem) 306: 77
72. Santosh C, Mishra PC (1994) Int J Quant Chem 51: 335

Using Molecular Electrostatic Potential Maps for Similarity Studies

Gyula Tasi[1,*] and István Pálinkó[2]

[1] Applied Chemistry Department, József Attila University, Rerrich B. tér 1., H-6720 Szeged, Hungary
[2] Department of Organic Chemistry, József Attila University, Dóm tér 8., H-6720 Szeged, Hungary

Table of Contents

1 Introduction 46

2 Molecular Electrostatic Potential: Performance of Semi-Empirical
  MO Calculations 46

3 Simplified Models for Calculating Molecular Electrostatic
  Potential Maps 53
  3.1 Point-Charge Models 53
  3.2 Multipole Expansion Methods 61

4 Molecular Electrostatic Potential Similarity Indices 62

5 Software Packages for Molecular Similarity Studies 68

References 69

In this work the use of molecular electrostatic potential (MEP) maps for similarity studies is reviewed in light of the latest results. First, methods of obtaining MEP maps is overviewed. The methodology, reliability and the efficiency of calculations based on semi-empirical as well as ab initio methods are discussed in detail. Point-charge models and multipole expansion methods which provide MEP maps of satisfactory quality are evaluated critically. Later on, similarity indices of various kinds are analyzed, compared and examples of their use are shown. Finally, the last section lists and summarizes software packages capable of calculating MEP map based similarity indices.

Topics in Current Chemistry, Vol. 174
© Springer-Verlag Berlin Heidelberg 1995

1 Introduction

Similarity is one of the fundamental concepts in chemistry as well as in biochemistry. Recently, several monographs dealt with a special kind of similarity, with molecular similarity [1, 2]. The concept of molecular similarity makes it possible to compare and classify the isolated molecules based on their individual properties, such as molecular geometry, dipole moment, charge distribution, etc. From the mathematical point of view, the relation that "two molecules are similar to each other by some of their properties" is an equivalency relation on the set of isolated molecules.

Quantum chemistry allows more or less reliable calculation of individual molecular properties (reliability very much depends on the approximations used). Thus, the results of quantum chemical calculations can be used for similarity studies directly. Researchers generally try to characterize the extent of molecular similarity with a real number, and therefore several scalar functions are defined as molecular similarity indices.

The electrostatic forces play a crucial role [3] in molecular interactions due to their long-range character. As a first approximation, the electrostatic interaction can be described by the electrostatic potential maps of the interacting isolated molecules. Obviously, the term of molecular interaction also includes host-guest interactions which are considered of the utmost importance in biological systems (enzyme-substrate, antigen-antibody, receptor-hormone, receptor-drug, etc.). In most instances the accurate structure of the host molecule is not known, and therefore one tries to learn about the structure of the binding site through the properties of the guest molecule [4]. In most cases molecules of different structure are capable of binding to the same site (of course with various strengths) giving various physiological responses. If the correlation between the physiological responses and the molecular similarity indices of the guest molecules is good, information of fundamental importance can be obtained about the properties of the host as well as the ideal guest molecules (here, ideal means the specific guest molecule we need).

We attempt here to describe those results which concern the quality of molecular electrostatic potential maps, methods of their calculation and their use in molecular similarity studies.

2 Molecular Electrostatic Potential: Performance of Semi-Empirical MO Calculations

The molecular electrostatic potential (MEP) is a rigorously defined quantum mechanical property. The electrostatic potential (EP) at a point r in the surroundings of a molecule is given in atomic units as follows:

$$V(r) = \sum_A^N \frac{Z_A}{||r - R_A||} - \int \frac{\varrho(r')\,dr'}{||r - r'||} \tag{1}$$

where N is the number of atoms in the molecule, Z_A is the charge of nucleus A located at point R_A and $\rho(r)$ is the electron density function of the molecule. The first term on the right-hand side of Eq. (1) is the classical contribution of the atomic nuclei, treated as point charges, to the EP. The second one gives the contribution of the electrons. It is to be seen that the two terms have opposite signs, and therefore their effects are opposite.

The EP defined by Eq. (1) can also be interpreted by the perturbation theory. The perturbation expansion of the energy $E(r)$ of a system consisting of a molecule interacting with a point charge Q located at point r is given as follows:

$$E(r) = E_0 + QV(r) + O^{(n)}(r) \tag{2}$$

where E_0 is the energy of the unperturbed (non-interacting) molecule and $QV(r)$ is the first-order energy term. Terms of higher order correspond to the polarization energy. When $Q = 1$ the first-order term numerically (disregarding the dimension) equals $V(r)$. Based on this interpretation EP is usually taken as an energy quantity.

Within the usual LCAO approximation, Eq. (1) transforms as follows:

$$V(r) = \sum_A^N \frac{Z_A}{||r - R_A||} - \sum_\mu \sum_\nu P_{\mu\nu} \int \frac{\varphi_\mu(r')\,\varphi_\nu(r')\,dr'}{||r - r'||} \tag{3}$$

where $P_{\mu\nu}$ is an element of the first-order density matrix P, φ_μ and φ_ν are atomic basis functions.

Since the EP is the expectation value of a one-electron operator r^{-1}, its calculation is correct to one order higher than the wavefunction used [5]. This means that the quality of the Hartree-Fock (HF) SCF wavefunction is generally appropriate for calculating EP when the molecule is in ground electronic state. For excited molecules correlated wavefunction is necessary for EP calculations [6].

Weinstein et al. [7] showed that for atomic systems the EP does not possess local maxima. Sen and Politzer [8] proved that for a monatomic negative ion the EP does have a unique minimum at a radial distance $r_m < \infty$. This work has led to several interesting studies on molecular negative ions [9–11]. For molecular systems Pathak and Gadre showed that the MEP map does not have local maxima [9]. Silberbach corrected their proof mathematically [10]. It also follows that when there are two or more local minima on the MEP map, they are connected with regions with saddle points. In Fig. 1 the contour plot of HF/4-31G MEP of benzene in its plane is displayed. Points P and P' in the figure are saddle points. Pathak and Gadre showed that for a negative molecular ion at least one local minimum with negative value must exist [9]. Gadre et al. have written a computer code determing the critical points of MEP maps [11].

The use of MEP maps in various areas of science has been the subject of several excellent reviews [5, 12–15].

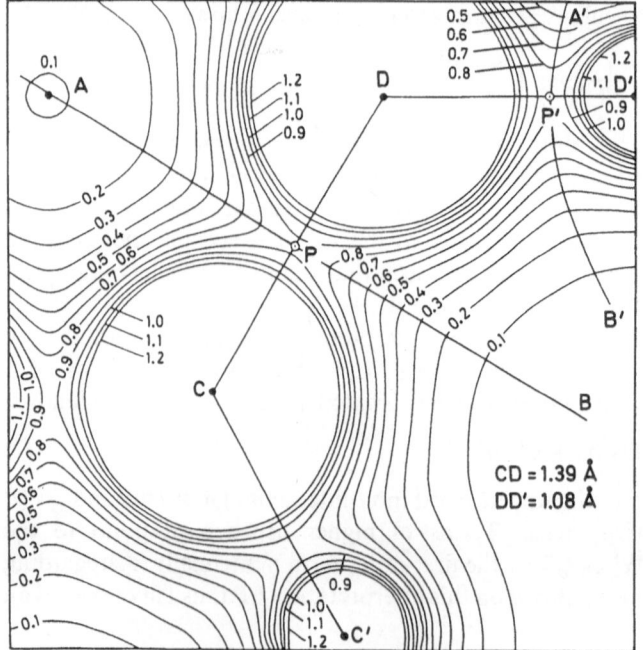

Fig. 1. Molecular electrostatic potential contour plot (values in au) for a portion of the benzene molecule in the molecular plane. Carbon nuclei are located at C and D with the hydrogen nuclei at C' and D'. Saddlepoint-like structures are seen at the points P and P'. (Reproduced from [9]; copyright-American Institute of Physics)

In common applications EP values must be calculated over several thousand points, therefore the efficiency of the computation technique is of crucial importance. Calculation of the first term on the right-hand side of Eq. (3) is trivial, but not of the second term because of the three-center one-electron potential integrals.

The molecular electron density function needed for EP calculation can be obtained through ab initio as well as various semi-empirical methods. Since ab initio calculations are not economical for large molecules (several hundred atoms), the use of well-parameterized semi-empirical methods are still justified. When semi-empirical methods are used the three-center potential integrals usually disappear, and therefore the electronic contribution can be easily calculated by Slater-type orbitals. In ab initio methods (primitive or contracted) Gaussian-type orbitals are used for calculating the three-center integrals because their calculations are clumsy with Slater-type orbitals.

Recently, Pople et al. [16] and Gadre et al. [17, 18] have realized nearly two orders of magnitude increase in the speed of ab initio EP calculations using the GAUSSIAN 92 [19] and INDMOL codes, respectively. This achievement will certainly open new areas for ab initio applications in the near future. Pople et al. tested the effectivity of their method on a large oligonucleotide molecule

(CTCGAG, $C_{116}H_{138}N_{46}O_{68}P_{10}^{-10}$) as well [16]. For this molecule the HF/STO-3G EP values were calculated over 9310 points. The CPU time was only 163 s on a Cray Y-MG/8-32 computer. This is negligible compared with the CPU time (11 h) necessary for the SCF calculation.

Using semi-empirical instead of ab initio methods for calculating molecular electron density functions and MEP maps speeds up calculations considerably. Early CNDO [20] and INDO [21] studies revealed that these methods, keeping the ZDO approximation, were not suitable for preparing MEP maps [22]. There was better agreement with ab initio and experimental results if after the SCF calculation a quasi ab initio electron density function was prepared by inverse Löwdin's transformation (deorthogonalization) [23] from the molecular electron density function calculated within the ZDO approximation [24] and retaining the three-center integrals in the EP calculation (quasi ab initio method) [22]. Thus, in this case, the method used for calculating MEP maps only differs from the minimal basis ab initio method in (i) that the inner electrons are attached to the nuclei, and consequently effective core charges are used for calculating the first term of Eq. (3) and (ii) that the molecular electron density function applied was prepared from the semi-empirical ZDO electron density function by deorthogonalization of the atomic basis functions. However, this method was unable to reproduce the main characteristics of the ab initio MEP maps for aromatic systems, i.e. the negative regions above and below the aromatic rings [25]. Culberson and Zerner showed that the INDO/S method [26], using the above mentioned deorthogonalization procedure, was able to reproduce the negative regions above and below the plane of the benzene molecule [27]. For the cytosine molecule the INDO/S method indicated that the most negative region was around the oxygen atom (O_2) [27]. This contradicts the conclusions derived from the HF/STO-3G MEP map [28] (see Fig. 2) and proton affinities [29, 30] which indicate that the molecule is protonated at N_3. It should be mentioned that the HF/4-31G [27] and the HF/6-31G* MEP maps [31] also favor the oxygen atom.

Recently, the MNDO type methods (MNDO [32], AM1 [33] and PM3 [34]) have been tested for their ability to produce reliable MEP maps. These semi-empirical methods, just as the CNDO and INDO methods, are ZDO methods, and are based on the more sophisticated NDDO approximation [35].

Illas et al. compared in detail the MEP maps calculated by ab initio method using a variety of Gaussian basis sets [36] with those obtained by the MNDO method [37]. Since the MNDO method is based on the ZDO approximation, in accordance with the previously published results [22, 27], the ZDO molecular electron density function was used for generating the MEP maps after inverse Löwdin's transformation (deorthogonalization). It was found that, whereas the MNDO MEP maps could reproduce the main characteristics of the HF/6-31G* MEP maps, the position of the MNDO minima were too close to the molecules and they were deeper than the corresponding ab initio ones. Fortunately, the ratios of the HF/6-31G* and MNDO MEP minima energies are constant, and therefore the MNDO MEP minima energies can be scaled to reproduce the

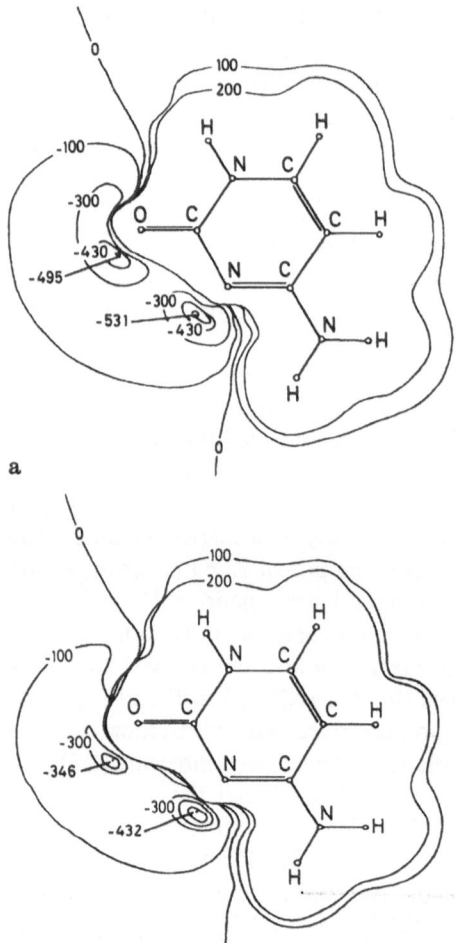

a

b

Fig. 2. Electrostatic potential maps for cytosine in the ring plane (kJ/mol): **a** calculated with *AM*1; **b** calculated with STO-3G. (Reproduced from [40]; copyright-John Wiley & Sons)

HF/6-31G* MEP minima energies. It was also found that the MNDO MEP maps would give the negative regions above and below the ring in aromatic systems. Moreover, in molecules such as cytosine, where there are more negative regions, the sequence of the MEP minima were also correctly reproduced [37]. Luque and Orozco have, in detail, studied the applicability of the AM1 method for calculating MEP maps [38]. Similar to the MNDO calculations, the AM1 wavefunction was used after deorthogonalization. They found that the positions of the AM1 MEP minima, just as those of the MNDO ones, were closer to the molecules than those calculated with ab initio methods. Calculations revealed that the AM1 method produced MEP maps similar to those of the HF/6-31G* method for certain molecules (e.g. heteroaromatic systems), while the agreement was not satisfactory for others (e.g. HCN). Orozco et al. studied the PM3 method in preparing MEP maps [39]. In these studies the PM3 method was found to be the most suitable among the MNDO type methods for calculating minima in the MEP maps. However, it should not be applied to the water molecule and its derivatives. Calculations revealed that the AM1 method could be recommended for studying the lone pairs of heteroaromatic systems, while MEP minima corresponding to π-electron charge distributions could be best studied with the MNDO method [39].

In several works the NDDO approximation was maintained, i.e. the original ZDO MNDO type wavefunction was used without deorthogoanlization for calculating MEP maps [40–44]. Within the NDDO approximation Eq. (3) is modified as follows:

$$V(r) = \sum_A^N \frac{Z_A}{||r - R_A||} - \sum_A^N \sum_{\mu, \nu \in A} P_{\mu\nu} \int \frac{\varphi_\mu(r') \varphi_\nu(r') \mathrm{d}r'}{||r - r'||}. \tag{4}$$

Thus, the three-center potential integrals are not retained. Ferenczy et al. found that the AM1 MEP maps are able to reproduce the main characteristics of the HF/STO-3G MEP maps for the water, formaldehyde, formamide and the cytosine molecules and for the cyanate ion [40]. However, the NDDO AM1 MEP maps gave deeper minima which were closer to the molecules than those in the HF/STO-3G MEP maps. The contour plots of MEP for the cytosine molecule are displayed in Fig. 2. It can be seen that the NDDO AM1 MEP map correctly predicts the N_3 nitrogen atom as a primary protonation center instead of the O_2 oxygen atom. This finding is in agreement with the HF/STO-3G MEP map [28] and the experimental as well as the theoretical proton affinities [29, 30]. Similar results were also obtained by Luque et al. based on the quasi ab initio MNDO MEP map [37]. INDO/S, HF/4-31G and HF/6-31G* calculations showed an opposite order of protonation [27, 31] as discussed earlier.

It is important whether the NDDO MNDO or AM1 MEP maps are able to reproduce the negative regions above and below the plane of the ring in aromatic systems. Reynolds et al. showed that the NDDO MNDO or AM1 MEP maps contained negative regions when 3-methyl-5-hydroxy- and 3-methyl-6-hydroxyindoles were used as models [41]. The MEP maps calculated by MINDO/3 method [45] (which is based on the INDO approximation) did not reveal such characteristics [41]. In Figs. 3 and 4 the contour plots of the NDDO AM1 MEP of the benzene molecule are displayed. The planes considered are respectively 1.75 Å (van der Waals radius of the C atom) and 1.24 Å above the plane of the molecule. The negative region above the ring can be readily observed. The AM1 wavefunction on the AM1 geometry and the contour plots of NDDO AM1 MEP of the benzene molecule were obtained with the PcMol package [46]. It seems worthwhile to compare the positions and the values of the minima with AM1 results obtained by Luque and Orozco [38]. As a reminder, they prepared a quasi ab initio electron density function with the help of inverse Löwdin's transformation and they kept the three-center potential integrals in Eq. (3). The calculated values are as follows: the minimum energy is − 40.98 kcal/mol and they lie above and below the plane of the ring and their distance from the center of the ring is 1.23 Å [38]. Our NDDO AM1 EP

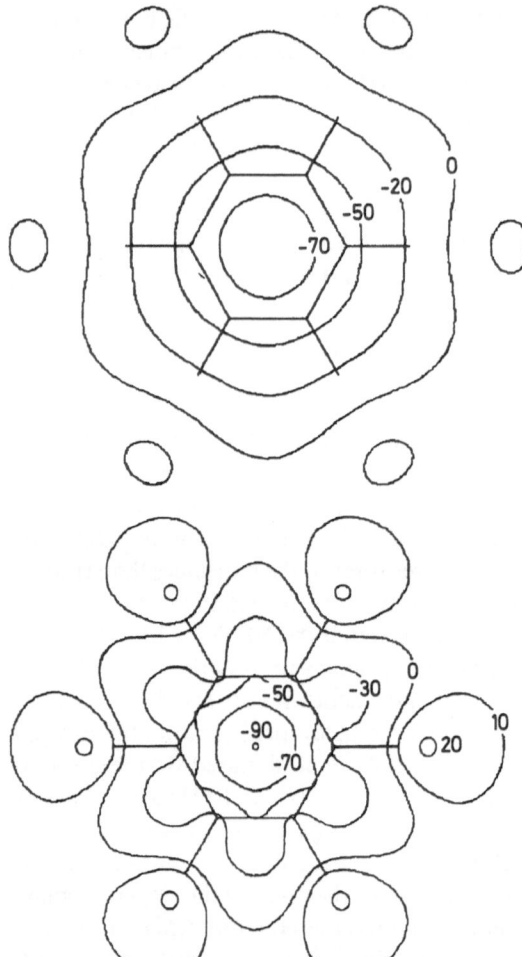

Fig. 3. The contour plot of the NDDO *AM*1 *MEP* of the benzene molecule 1.75 Å above the molecular plane (kJ/mol)

Fig. 4. The contour plot of the NDDO *AM*1 *MEP* of the benzene molecule 1.24 Å above the molecular plane (kJ/mol)

calculations, for the value and the position of the minima, determined by the simplex method [47], gave − 21.55 kcal/mol and 1.24 Å, respectively and the negative regions were above and below the plane of the ring. The results obtained by HF/6-31G* EP calculations are − 20.52 kcal/mol and 1.76 Å, respectively [38]. It can be seen that the NDDO AM1 MEP minimum energy is closer to the HF/6-31G* MEP minimum energy than that computed by the quasi ab initio method, however, both methods giving the positions of the minima too close to the center of the ring. The above detailed results establish that the ZDO AM1 and MNDO MEP maps are of acceptable quality: they are able to reproduce the main characteristics of the ab initio MEP maps [41].

Let us compare the SCF dipole moments obtained by MNDO type methods either from the NDDO or the quasi ab initio wavefunctions to the experimental dipole moments. Since experimental results were taken into account in the parametrization of MNDO type methods [32–34], it is more appropriate to compare them to experimental dipole moments instead of HF ab initio ones. The experimental dipole moments of the molecules selected for the parametrization procedure were also taken into account with a certain weight in the least-squares optimization method. The philosophy of the parametrization process thus differs from that of the CNDO/2 method [20] where minimum ab initio methods served as reference. The results of our calculations [42, 43] in accordance with the parametrization philosophy of MNDO type methods prove that the SCF dipole moments obtained from NDDO wavefunctions were closer to the experimental dipole moments than those calculated from quasi ab initio wavefunctions.

Cummins and Gready also maintained the NDDO approximation for the calculation of MNDO and AM1 MEP maps [48]. The EP was taken as the first-order term of the perturbation expansion of the interaction energy and the unit positive test charge was regarded as a hydrogen nucleus. Eventually the contribution of the atomic nuclei was computed with the help of core-core repulsion functions used in the semi-empirical method in question. For computing the contribution of electrons the Goepert-Mayer-Sklar potential was applied with the approximation of neglecting the penetration integrals. This potential is generally used for calculating the core-electron attraction integrals in the MNDO type methods [32]. Ford and Wang calculated the electronic contribution similarly [31]. However, for evaluating the contribution of atomic nuclei, they applied a semi-empirical potential function containing two parameters per atom. These parameters were determined from the HF/6-31G* MEP maps of some small molecules. Their method is particularly suitable for investigating the behavior of heteroatoms with lone electron pairs and for π-systems [31].

Beside the above mentioned semi-empirical methods, the local density functional theory [49] gives an alternative possibility for calculating MEP maps [50]. The MEP maps for benzene, ethylene, formamide, cytosine and 2,3,7,8-tetrachlorodibenzo-p-dioxin computed with DNP (double numerical plus polarization) basis set were able to reproduce the main characteristics of the ab initio MEP maps [50]. This method seems to be very suitable especially for large molecules.

3 Simplified Models for Calculating Molecular Electrostatic Potential Maps

3.1 Point-Charge Models

The representation of molecules by point charges, mostly with monopoles at the atomic sites (atomic charges), has a long history in chemistry and the concept proved to be very useful [51]. The atomic charges are used extensively, e.g., in characterization of reactivity of various atomic sites in molecules, representing molecular dipole moments, for calculating electrostatic interaction in molecular mechanics and dynamics and for considering solvation effects in quantum chemical calculations [52–54]. For large biomolecules, atomic charges provide one of the very few ways of preparing MEP maps.

Recently, several excellent reviews have been published concerning the representation of molecules by point charges [55–57], and therefore we restrict ourselves to discussing the latest results.

Unfortunately there is no unique quantum mechanical recipe for calculating atomic charges: mostly one has to apply some sort of scheme for the generation of the classical point-charge system. However, one must be able to evaluate the quality of the atomic charges. A possible quality characteristic can be the agreement of the dipole moment of the classical point-charge system with the quantum chemical or the experimental dipole moment. Of course, higher multipole moments can also be used if they are accessible. For larger, asymmetrical molecules, however, this comparison does not give useful information. A better quality characteristic can be how well the point charge system is able to reproduce the electrostatic field in the surroundings of the molecule. Since the MEP is a strictly determined quantum mechanical property it provides a good way for evaluating the quality of the classical point-charge system.

In a plausible manner atomic charges can be derived from quantum mechanical MEP maps through constrained optimization [6, 39–44, 58–66]. These atomic charges are called EP or PD (Potential Derived) charges. Usually only one constraint is taken into account: the sum of atomic charges ought to equal the charge of the molecule. This constraint can be fulfilled most easily by using the Lagrange multiplier method [67]. (Other constraints can also be taken into account, e.g., for the dipole moment vector of the atomic charges.)

For the computation of EP charges, points are first generated outside the van der Waals surface of the molecule by using an appropriate algorithm and trial monopoles are placed at the sites of atoms and the classical electrostatic field generated by them is fitted to that obtained quantum chemically using the least-squares methods. The object function is given by

$$S(q_1, q_2, \ldots, q_n, \lambda) = \sum_i^M \left[V_i - \sum_j^N \frac{q_j}{r_{ij}} \right]^2 + \lambda \left[\sum_j^N q_j - q_{\text{mol}} \right], \tag{5}$$

where M is the number of points generated outside the van der Waals surface of the molecule, N is the number of atoms, V_i is the quantum chemical EP at point i, r_{ij} is the distance between point i and atom j, q_j is the charge of the atom j, q_{mol} is the charge of the molecule and λ is the Lagrange multiplier. After elementary mathematical operations a matrix equation is obtained. Finally, for calculating the atomic charges, only one matrix inversion must be performed [57]. The remaining task is just checking the regularity of the matrix. The goodness of fit can be characterized by the root mean square deviation (rmsd):

$$\text{rmsd} = \left(\frac{\sum_i^M (V_i^{el} - V_i^{qm})^2}{M} \right)^{1/2} \tag{6}$$

where V_i^{qm} is the quantum chemical EP and V_i^{cl} is the classical EP at point i. The relative rmsd (rmsd_{rel}) can also be used:

$$\text{rmsd}_{\text{rel}} = \left(\frac{\sum_i^M (V_i^{cl} - V_i^{qm})^2}{\sum_i^M (V_i^{qm})^2} \right)^{1/2} \tag{7}$$

These atomic charges are usually capable of reproducing the calculated dipole moment and the quantum chemical MEP map of the molecule outside the van der Waals surface. The MEP map of the classical EP system is not so rich in information, however, and there are problems, e.g., with the lone pairs. The local anisotropies due to the lone pairs largely disappear. In order to obtain a more accurate description further point charges, situated in appropriate sites, are usually taken into account during optimization [68].

The quality of the EP charges basically depends on the quality of EP. For the calculation of EP each method described in Sect. 2 can be used. Thus the EP charges can be calculated by semiempirical as well as ab initio methods.

Studying the MNDO type methods (MNDO, AM1 and PM3) in preparing EP charges from quasi ab initio density matrix led to the conclusion that the MNDO method is best while the PM3 method is worst for this purpose [39, 64]. After scaling the quasi ab initio EP charges computed by the MNDO method are able to reproduce the EP charges derived from the HF/6-31G* MEP maps [64]. The semi-empirical quantum chemical methods concentrate the charge density closer to the atomic nuclei than the ab initio methods, and therefore the absolute value of the semi-empirical MEP is smaller at a longer distance from the molecule than the ab initio one. That is why the scaling factor is higher than 1.0.

Choosing an appropriate method for generating points outside the van der Waals surface of the molecule and the proper density of points are also important problems to solve. To do it correctly, one must take into account the basic characteristics of the MEP maps. The magnitude of the EP and the complexity of the MEP maps decrease approximately radially from the molecule. For illustration, see Fig. 5.

It is obvious that a uniform cubic grid of points is not a good choice since in this case many points are required for the adequate description of the MEP map. For uniform cubic grid there will be too many points at large distances where the MEP maps are poor in information. As a consequence, an immense number of points would be needed to obtain stable EP charges [69].

·An alternative possibility is using concentric extended van der Waals surfaces and choosing points on these shells in some way. Kollman and Singh [60] applied Conolly's molecular surface algorithm [70]. Chirlian and Francl [61] generated concentric shells on the atoms, each shell containing 14 points with practically uniform distribution, and retained those points which were not in the extended van der Waals spheres of the other atoms (CHELP program). However, the original CHELP routine was not reliable for conformational studies [69].

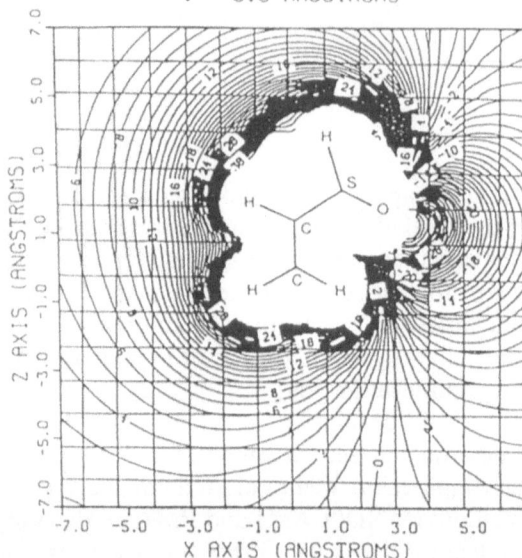

Fig. 5. A contour plot representing a two-dimensional slice of the molecular electrostatic potential for vinyl sulfone in the plane of the molecule with a sample uniform cubic grid superimposed; grid points would be placed at the intersections of the perpendicular lines. (Reproduced from [71]; copyright-John Wiley & Sons)

A further possibility is the MACRA (Molecular Atom-Centered RAdial) algorithm [71], in which a unit sphere is generated with approximately uniform distribution of points and then this template sphere is used for preparing the concentric shells. This template sphere is placed at the atomic centers and scaled in the appropriate manner. For the first shell $R(1) = 2$ Å is the suggested radius, while for the others it is $R(I) = R(I - 1) + GS(I - 1)$ $(I \geqslant 2)$, where GS is the point spacing of the shell in question, the average distance between neighboring points. Only those points which do not fall in the sphere of the other atoms are retained. As a result, the density of points decreases radially from the molecule. A schematic two-dimensional MACRA grid is displayed in Fig. 6. Using this algorithm the number of points required for storing and representing the information of the MEP map can be decreased significantly.

Lee and Friesner suggested an approximation for calculating EP charges for large molecules (about 1000 atoms) [72]. Since the preparation of a first-order density matrix used for computing ab initio EP charges for large molecules requires considerable CPU time even when minimum basis set is applied, it is worthwhile preparing the first-order density matrix of the whole molecule from the scaled elements of the first-order density matrices of the appropriately chosen fragment molecules and setting the off-diagonal elements to be zero after a given neighborhood (fragment density matrix approach) [72].

For large molecules Orozco and Luque also suggested a practical procedure to calculate EP atomic charges [73]. They divided the molecule into several parts and these parts were treated independently concerning the calculation of EP atomic charges (fractional model). The influence exerted by the other parts is taken into account with the help of small groups of atoms.

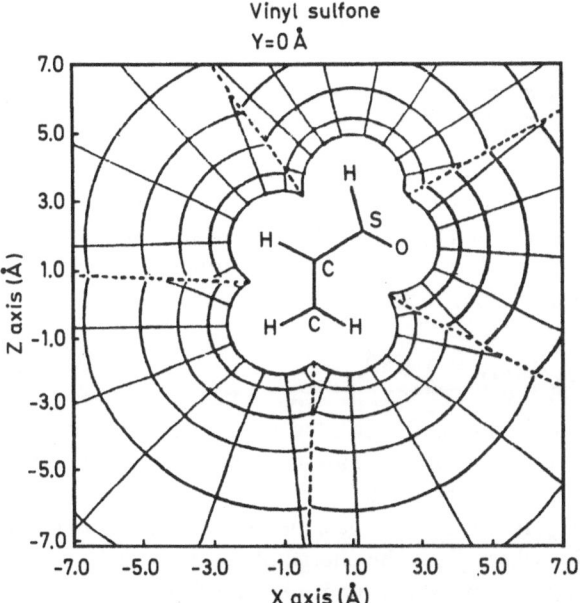

Fig. 6. A two-dimensional slice of sample MACRA grid, where grid points correspond to the intersections of the radial lines and circular spheres. (Reproduced from [71]; copyright – John Wiley & Sons)

For lowering the CPU time of SCF calculations for large molecules, the fragment SCF method [74] can also be used.

In order to determine atomic charges, Su performed a least-squares fit to the EP values calculated at sites of atomic nuclei [75]. Thus, the approximating character of fitting on grid points outside the van der Waals surface of the molecule was eliminated. For calculating atomic charges, electron density functions derived from XRD data [76] were applied.

There are several computer programs available for calculating EP charges [46, 60, 61, 75, 77, 78].

Since the EP charges are determined through parameter estimation from exact quantum chemical MEP maps, it is not surprising that they are able to reproduce the electrostatic field outside the van der Waals surface of the molecule. There are several other methods for calculating atomic charges which are computationally simpler and do not require much CPU time. Due to their simplicity, the Mulliken and Löwdin population analysis methods [79, 80] have long been used for calculating atomic charges. These methods are included in standard quantum chemical packages. Unfortunately, however, in most cases atomic charges obtained through these methods cannot reproduce either the dipole moments of the molecule or (consequently) the molecular electrostatic field.

The main features of the Mulliken population analysis are as follows [79, 81]. First, the atomic and overlap electron population are computed with the help of the LCAO-MO SCF wavefunction. The overlap electron population Q_{AB} between atoms A and B can be determined with the following equation ($A \neq B$):

$$Q_{AB} = \int \rho_{AB}(r)dr = 2 \sum_{\mu \in A} \sum_{\nu \in B} P_{\mu\nu}S_{\mu\nu} \qquad (8)$$

where $\rho_{AB}(r)$ is the electron density function corresponding to atoms A and B, $P_{\mu\nu}$ and $S_{\mu\nu}$ is an element of the first-order electron density matrix P and the overlap integral matrix S, respectively. The first approximation in the analysis is that the overlap electron population Q_{AB} is shared equally between atoms A and B. Thus, the gross electron population of atom A is:

$$Q_A = \sum_{\mu \in A} P_{\mu\mu} + \sum_{\mu \in A} \sum_{\nu \in B} P_{\mu\nu}S_{\mu\nu} = \sum_{\mu \in A} (PS)_{\mu\mu}. \qquad (9)$$

It is to be seen that the trace of the matrix PS gives the number of electrons in the molecule. The second approximation in the analysis is that the total electron population given by Eq. (9) is placed on atom A, i.e. the position of atom A is assumed to be the center of Q_A. In this way one gets the Mulliken (MP) charge of atom A: $q(MP)_A = Z_A - Q_A$, where Z_A is the charge of the nucleus A.

The latter of these two approximations cannot be justified. Through using second quantization formalism [82] Mayer showed that halving is a consequence of the LCAO approximation [83]. However, it does not mean that the second approximation is valid and the MP charges are the true atomic charges within the LCAO approximation.

Huzinaga et al. modified the Mulliken population analysis: the overlap electron population is not halved, the centers of the electron populations are computed, and thus the arbitrary assumptions are eliminated [81]. The center of the electron population Q_{AB} is given by the following equation:

$$\langle \alpha \rangle_{AB} = \frac{\int \alpha \varrho_{AB}(r)dr}{Q_{AB}} = \frac{\sum_{\mu \in A} \sum_{\mu \in B} P_{\mu\nu} \langle \mu | \alpha | \nu \rangle}{\sum_{\mu \in A} \sum_{\mu \in B} P_{\mu\nu}S_{\mu\nu}} \qquad (10)$$

where $\alpha = x,y,z$. With the help of this extended Mulliken population analysis the molecules can be represented by a classical point-charge system, the components of which are the atomic nuclei, the atomic electron populations and the overlap electron populations. For the ab initio or quasi ab initio LCAO-MO SCF electron density function, a molecule consisting of N atoms can be represented by a point-charge system containing $N + N(N + 1)/2$ charges. The same molecule can be described by 2N charges with the NDDO approximation. The extended Mulliken point-charge system is able to reproduce the SCF dipole moment of the molecule quite well [81].

For reducing the number of point charges, i.e. calculating atomic charges, the same method can be used as for calculating EP atomic charges, but in Eq. (5) the exact quantum chemical EP has to be replaced with the classical EP generated by the extended Mulliken point-charge system [42, 43]. In this way the simplicity of electron population analysis methods and the unambiguous character of atomic charges derived from quantum chemical MEP maps can be combined. The atomic charges obtained are called population electrostatic potential (PEP) atomic charges and can be calculated from any LCAO-MO SCF wavefunction. Using the MNDO and AM1 methods Tasi et al. showed that the PEP atomic charges are able to reproduce the SCF dipole moments of molecules just as well as the EP charges derived from exact quantum chemical MEP maps. Moreover, these calculations require considerably less CPU time [42, 43]. In Table I the results of the μ(method) = $\beta^*\mu$(SCF) linear regressions for 41 molecules are listed [43]. Using the MNDO, AM1 and PM3 methods Tasi et al. also concluded that for nucleotides the PEP atomic charges reproduce the NDDO EP charges and thus the electrostatic field outside the van der Waals surface quite well [44]. Of course this means that the initial extended Löwdin point-charge system is also suitable for the preparation of MEP maps outside the molecular van der Waals surface. In Figs. 7 and 8 the relations q(EP) vs q(MP) and *q(EP)* vs q(PEP)— are displayed, obtained for nucleotide bases using the PM3 method [44].

Rauhut and Clark used the AM1 method within the NDDO approximation for calculating atomic charges [84]. The point charges corresponding to the

Table 1. Statistical data obtained by linear regression of the dipole moments. (Reproduced from [43]; copyright– John Wiley & Sons)

Method	NDDO approximation			After deorthogonalization		
	β	$SE(\beta)$	r	β	$SE(\beta)$	r
MP	0.8431	0.0291	0.9776	1.2243	0.0496	0.9695
PEP	1.0026	0.0039	0.9997	0.9974	0.0033	0.9998
EP	1.0099	0.0026	0.9999	1.0160	0.0036	0.9998

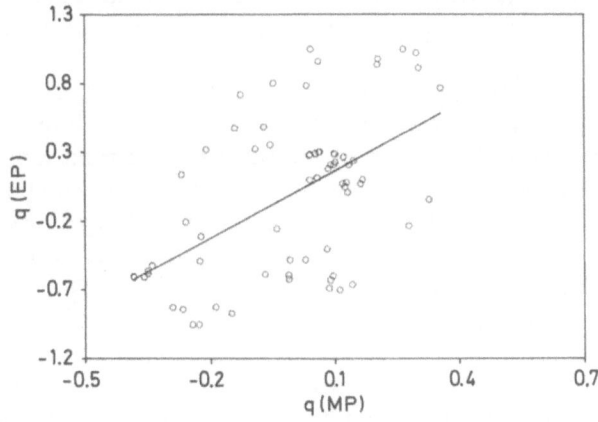

Fig. 7. Graphical representation of the relation $q(EP)$-*PM*3 vs $q(MP)$-*PM*3 for nucleotide bases. (Reproduced from [44]; copyright-American Chemical Society)

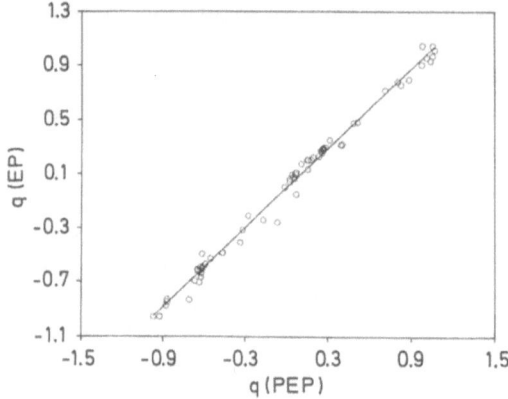

Fig. 8. Graphical representation of the relation $q(EP)\text{-}PM3$ vs $q(PEP)\text{-}PM3$ for nucleotide bases. (Reproduced from [44]; copyright-American Chemical Society)

electron distribution were situated at the center of charge density of each lobe of the AM1 natural atomic orbitals (NAOs) [85]. In this way each heavy atom is represented by a maximum of nine point charges (the atomic center and two centers per NAO). This point-charge system was used for computing MEP maps and, from the results obtained on a relatively small set of molecules, it was concluded that the properties of the HF/6-31G* MEP maps were reproduced well. When quasi ab initio AM1 EP charges [63] were calculated by the MOPAC package modified by Merz and Besler [78] the agreement was not so good [84].

Ferenczy [86] derived atomic charges from the EP of multipole moments (DMMs) obtained from a distribution multipole analysis (DMA) [87]. In this method, instead of calculating the EP in points of a relatively arbitrary grid, a definite integral must be calculated analytically over a certain region of the space around the molecule. Atomic charges obtained give the best reproduction (in the least-squares sense) of DMMs (DMM charges). This method is faster than comparing to computing EP charges and the point charges reflect the molecular symmetry properties and do not depend on the orientation of the molecule. The NDDO AM1 DMM atomic charges are able to reproduce the NDDO AM1 EP charges [86]. Scheraga et al. applied Ferenczy's method [86] for the ab initio (HF/6-31G, 6-31G*, 6-31G**) wavefunctions of saturated hydrocarbon molecules [88]. It was concluded that the DMM atomic charges, in contrast with the EP atomic charges, were transferable and show a consistent $C^{\delta+} - H^{\delta-}$ polarity. The EP charges are only transferable if off-atomic point charges (e.g. on the bonds) are also taken into account. The transferability of atomic charges are discussed in several papers as well [66].

Recently, Sokalski et al. presented distributed point charge models (PCM) for some small molecules, which were derived from cumulative atomic multipole moments (CAMMs) or from cumulative multicenter multipole moments (CMMMs) [89, 90] (see Sect. 3.2). For this method the starting point can be any atomic charge system. In their procedure only analytical formulas are used, without involving an approximate fitting process applied for calculation of EP and PEP charges. In the literature several CAMM databases are available for amino acids and for nucleotide bases [91, 92]. Each CMMM (M_c^{klm}) can be expressed as a set of point charges q_p located at a position $p(x_p, y_p, z_p)$ at a distance R from the expansion center c:

$$M_c^{klm} = \sum_{p=1}^{s} q_p(x_p - x_c)^k (y_p - y_c)^1 (z_p - z_c)^m. \tag{11}$$

Charges can be obtained at different level of moments such as monopole ($s = 1$), dipole ($s = 3$) and quadrupole ($s = 9$). Torsion energy barriers for the HS-SH molecule calculated by several methods can be seen in Fig. 9 [90]. For the PCM model of this molecule the number of expansion centers is six ($c = 6$): beside the atomic centers, one center per S-H bond is further included. It can be seen that the PCM result is very close to the CMMM one and the PCM charges can be used for calculating intramolecular electrostatic interactions as well.

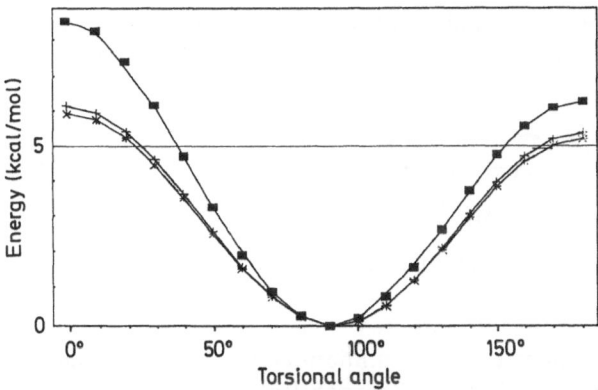

Fig. 9. Torsional potentials for HS-SH molecule calculated in 6-31G* basis set: (a) ab initio SCF results, ■ ab initio SCF; (b) CMMM estimates ($c = 6$) up to quadrupole-quadrupole term, + CMM (Q-Q); (c) PCM results ($c = 6$, $s = 9$, $R = 0.1$ au), *PCM ($R = 0.1$ au) (Reproduced from [90]; copyright-Springer-Verlag)

3.2 Multipole Expansion Methods

One of the advantages of the molecular (one center) and multicenter multipole expansions of the MEP is that, through truncating the series after some terms, one can get an analytical expression. The molecular multipole expansion, in contrast with the multicenter multipole expansion, diverges at distances of chemical interest. The various multicenter multipole expansion [87, 89, 93–96] practically equivalent to each other [97].

However, there are several problems with the multicenter multipole expansions as well, such as (i) how many centers are to be taken, (ii) where to place them and, finally (iii) how many terms (monopole, dipole, quadrupole, etc.) per center are to be taken into account.

The multipole moments of the molecular charge distribution are the expectation values of the operators $u^k v^l w^m (u,v,w = x,y,z)$:

$$\langle u^k v^l w^m \rangle = \sum_i^N Z_i u_i^k v_i^l w_i^m - \sum_{\mu,\nu} P_{\mu\nu} \langle \varphi_\mu | u^k v^l w^m | \varphi_\nu \rangle. \tag{12}$$

The first and second terms are the contributions of the atomic nuclei and the electrons, respectively. The calculation of the multipole integrals in the second term is simpler and faster than that of the potential integrals. After rearrangement of Eq. (12) a molecular multipole moment can be expressed as a sum of the atomic contributions $\langle u^k v^l w^m \rangle_i$:

$$\langle u^k v^l w^m \rangle_i = Z_i u_i^k v_i^l w_i^m - \sum_{\mu \in i} \sum_\nu P_{\mu\nu} \langle \varphi_\mu | u^k v^l w^m | \varphi_\nu \rangle. \tag{13}$$

These atomic contributions depend on the choice of the coordinate origin. A space invariant form can be obtained using the cumulative approach [93, 94]. The space invariant cumulative atomic multipole moments (CAMMs) do not contain contributions from lower moments. The definition of the CAMMs is as follows:

$$M_i^{klm} = \langle u^k v^l w^m \rangle_i - \sum_{k'>0}^k \sum_{l'>0}^l \sum_{m'>0}^m \binom{k}{k'}\binom{l}{l'}\binom{m}{m'}$$

$$\times u_i^{k-k'} v_i^{l-l'} w_i^{m-m'} M_i^{k'l'm'} \tag{14}$$

where $k'l'm' = klm$. The M_i^{000} is the atomic charge (monopole) at center i and it can be obtained from an appropriate atomic charge scheme (see Sect. 3.1). The ambiguous character of the atomic charges used can be corrected with the help of higher cumulative atomic moments. The multicenter multipole moment expansion can be further improved if additional expansion centers, not only on the atomic nuclei but also on the bonds, are taken into account [86, 95]. The terms of this expansion are called cumulative multicenter multipole moments (CMMs) [98].

Sawaryn and Sokalski prepared an interface to the Gaussian package which is capable of calculating various CAMMs/CMMMs and distributed point charges (PCM) for their representations [89, 90]. The MOL17 and PCMCAMM programs developed by Sokalski and Sneddon are suitable for displaying MEP maps [98].

4 Molecular Electrostatic Potential Similarity Indices

Carbo et al. introduced the concept of similarity index (SI) for measuring the similarity of two molecular electron density functions [99, 100]. The Carbo *SI* for molecules A and B is as follows:

$$SI_{(C)} = \frac{\int \varrho_A \varrho_B d\tau}{\left(\int \varrho_A^2 d\tau\right)^{1/2} \left(\int \varrho_B^2 d\tau\right)^{1/2}} \tag{15}$$

where ϱ_A and ϱ_B are the electron density functions for molecules A and B, respectively and the integration must be performed over all space. The numerator is a measure of the overlap of charge densities for two superimposed molecules. The denominator normalizes the SI. The value of $SI_{(C)}$ varies in the range 0 to 1. It has a value of 1 when the electron density distributions in the two molecules are identical. Since $SI_{(C)}$ depends on the relative position of the molecules to be compared, it is worthwhile to maximize (optimise) its value. During the optimization process the coordinates of one molecule are fixed (reference molecule) and the coordinates of the other molecule are changed (moving molecule). The $SI_{(C)}$ is a function of six parameters without involving intramolecular degrees of freedom: the vector components of the center of coordinates of the moving molecule relative to the reference molecule and the three rotation angles of the system axes of the moving molecular structure with respect to the reference [101]. Optimal value of $SI_{(C)}$ with respect to these six parameters must be determined. Since the calculation of $SI_{(C)}$ by ab initio methods is difficult [102], there were attempts to use simpler methods [101, 103]. The molecular electron density functions were approximated with Gaussian-type functions by Hodgkin and Richards [103], while Carbo and Calabuig used a CNDO-like approximation [101].

Hodgkin and Richards suggested a new SI to provide a more sensitive measure of similarity [104]:

$$SI_{(H)} = \frac{2\int \varrho_A \varrho_B d\tau}{\int \varrho_A^2 d\tau + \int \varrho_B^2 d\tau}. \tag{16}$$

The value of $SI_{(H)}$ also varies in the range 0 to 1. When $\rho_A = n\rho_B$ (n is constant and $\neq 0$) is substituted into Eq. (15), the Carbo index will be unity, which means that the Carbo index represents the similarity of the shapes of electron density distributions but not of the magnitudes. When $\rho_A = n\rho_B$ is substituted into Eq. (16), the result will be $SI_{(H)} = 2n/(1 + n^2)$, which means that the Hodgkin index characterizes the similarity in the shape as well as in the magnitude of the electron density functions [104].

For calculating the Carbo and the Hodgkin indices the molecular electro-static potential (MEP-SI) as well as the molecular electrostatic field (MEF-SI) can be used. The use of MEP and MEF functions is particularly attractive since they are more relevant for structure-activity studies than the molecular electron density function. Because of technical difficulties the MEP and MEF functions are calculated over a grid of points surrounding the superimposed molecules (grid-based MEP-SI/MEF-SI formalism). The MEP-$SI_{(H)}$ function was first used for comparing discrete grid MEP functions by Hodgkin and Richards

[104]:

$$MEP\text{-}SI_{(C)} = \frac{\sum_{i}^{M} V_{Ai} V_{Bi}}{\left(\sum_{i}^{M} V_{Ai}^2\right)^{1/2} \left(\sum_{i}^{M} V_{Bi}^2\right)^{1/2}} \tag{17}$$

where V_{Ai} and V_{Bi} are the EP values at point i generated by molecules A and B, respectively. Similarly, the Hodgkin $MEP\text{-}SI$ can be defined as follows:

$$MEP\text{-}SI_{(H)} = \frac{2\sum_{i}^{M} V_{Ai} V_{Bi}}{\sum_{i}^{M} V_{Ai}^2 + \sum_{i}^{M} V_{Bi}^2} \tag{18}$$

The values of $MEP\text{-}SI$s defined by Eqs. (17) and (18) vary in the range -1 to 1. For calculating $MEF\text{-}SI$s, the scalar products of the MEF vectors are taken into account.

Hodgkin and Richards studied the similarity of the bioisosteric molecules Me_2CH_2, Me_2O and Me_2S [104]. For calculating $MEP\text{-}SI$ and $MEF\text{-}SI$ values a uniform density cubic grid ranging from -10 Å to 10 Å along each Cartesian axis with grid points spaced 1 Å apart was used. Grid points within each molecular van der Waals surface were excluded. For calculating MEP values and MEF vectors, Mulliken atomic charges calculated by the MOPAC package [105] were used. In the Me_2CH_2/Me_2O and Me_2CH_2/Me_2S similarity investigations the Carbo $MEP\text{-}SI$ and $MEF\text{-}SI$ values were significantly higher than the values of the Hodgkin indices. For instance, in the case of Me_2CH_2/Me_2O similarity investigation the $MEP\text{-}SI_{(C)}$ and $MEP\text{-}SI_{(H)}$ values were 0.70 and 0.03, respectively. The interpretation of the result was that the pattern of atomic charges in each molecule was similar and consequently their MEPs and MEFs were similarly shaped. They concluded that the Hodgkin-type index was more sensitive to the differences in magnitude of MEP and MEF functions than the Carbo-type.

In a more recent work Richard [71] could not reproduce the results of Hodgkin and Richards [104]. Having difficulty reproducing the Mulliken atomic charges applied in [104], ab initio EP charges were used instead. From the results obtained using the same uniform density cubic grid as in the earlier work [104] Richard concluded that the $MEP\text{-}SI_{(C)}$ and $MEP\text{-}SI_{(H)}$ indices are equally discriminating for the bioisosteric molecules studied.

Richard modified the calculation method of SI indices as well [71]. Hodgkin and Richards used constant values for normalizing the SIs defined by Eqs. (17) and (18). Grid points within each molecular van der Waals surface are excluded [104]. Let $A0$ and $B0$ be these grid regions for molecules A and B, respectively. Let GA and GB be those regions which surround the molecules outside the van der Waals surface. Grid points in regions $GA \cap B0$ and $GB \cap A0$ do not contribute to the numerators of Eqs. (17) and (18) and therefore these regions should not be taken into account in calculating the denominators either [71]. Thus *SI* values will only depend on the relative coordinates of the molecules and will be independent of the shapes of regions *A*0 and *B*0.

Let us further discuss the reproduction problem in connection with the calculation of Hodgkin and Richards. The main problem was the irreproducibility of the Mulliken atomic charges obtained with the MOPAC package [105]. The failure was the result of several factors: (i) Hodgkin and Richards applied the MNDO method for calculating atomic charges, (ii) the old MNDO parameters [106] were used instead of the new ones for sulphur [107] and (iii) the Mulliken atomic charges were calculated within the ZDO approximation (Löwdin atomic charges). If all these factors are taken into account, the results of Hodgkin and Richards [104] proved to be reproducible.

Burt et al. studied the similarity of nitromethylene insecticides with the help of Carbo-type and Hodgkin-type *MEP*- and *MEF-SI*s [108]. For this work *MM*2 [109] and *AM*1 optimized molecular geometries were used. Mulliken and Löwdin atomic charges were calculated for both sets of molecular geometries using *AM*1, *MNDO* and ab initio STO-3G methods. These atomic charges were then used for computing *MEP* values and *MEF* vectors in grid points. The extent and the increment of the grid applied for evaluating the molecular SIs were optimized to reduce CPU time to a reasonable level. It was found that the *AM*1 geometries and the *AM*1 Mulliken atomic charges gave the most reliable results for the given set of molecules. For evaluating the molecular *SI*s the compound with the highest biological activity was chosen as a reference molecule (Fig. 10). The relation $MEP\text{-}SI_{(H)}$ vs biological activity is displayed in Fig. 11 [108]. (In structures designated by "O" the nitro and amino groups are on the opposite sides of the double bond). It can be seen that the correlation is good for most molecules.

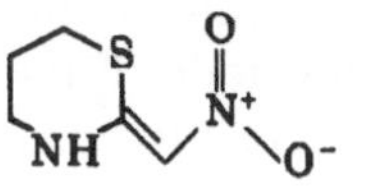

Fig. 10. The structure of the nitromethylene insecticide molecule (compound 185 O) used as a reference molecule in [104]

In calculating molecular SIs, Burt and Richards allowed intramolecular rotations about the torsional bonds of moving molecules [110]. Thus, the number of parameters in the object function was six plus the number of torsional rotations allowed. To exclude conformations with unreasonably large energies, they used a weighted SI in the optimization procedure:

$$\text{weighted } SI = \text{actual } SI_* \exp\left(\frac{-c\Delta E}{RT}\right) \qquad (19)$$

where c is the weighting factor, ΔE = energy of rotated conformation − energy of initial conformation, R is the gas constant and T is the temperature. For calculating ΔE an interaction potential containing an electrostatic and a van der Waals term was used. The necessary parameters were obtained from the *MM*2

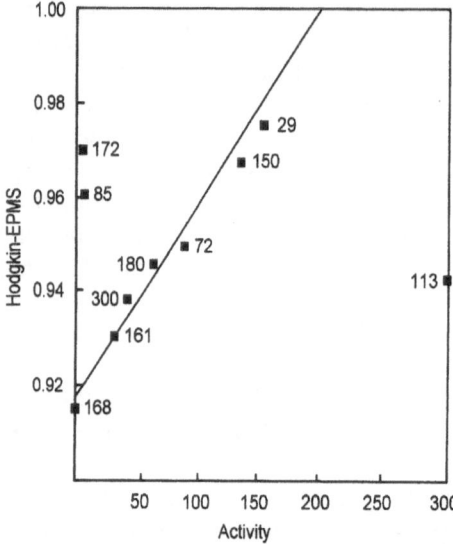

Fig. 11. Plot of the Hodgkin MEP-SIs (Hodgkin-EPMS) vs biological activity for the nitromethylene O structures (nitro and amino groups are on the opposite sides of the double bond) relative to compound 185 O (see Fig. 10) for the *AM*1 geometry/*AM*1 charge scheme. (Reproduced from [108]; copyright-John Wiley & Sons)

package [109]. *AM*1 optimized geometries and *AM*1 Mulliken atomic charges were used in the similarity study. The results were not satisfactory, however. One reason could be that identical atomic charges were applied for different conformations and the quality of the charges was not satisfactory either. Since atomic charges generally depend on the conformation of the molecule [111], the charge system for calculating *MEP* maps should be updated for significantly distinct conformations. Moreover, it would give the advantage of obtaining the quantum chemical energy of various conformers.

Sanz et al. also took into account the conformational flexibility of moving molecule in computing molecular *SI*s [112]. The energetically unfavorable conformers were excluded with the help of atomic van der Waals radii. If the distance of any two atoms in any conformer was smaller than 1.5 times the sum of their van der Waals radii, the conformer was excluded.

Good et al. used Gaussian-type functions for calculating *MEP-SI*s [113]. The *EP* of atomic charges at a point *r* in the surroundings of a molecule is as follows:

$$V^{el}(r) = \sum_{i}^{N} \frac{q_i}{||r - R_i||} \tag{20}$$

where q_i is the charge of an atom i located at R_i. Good et al. fitted a linear combination of two or three Gaussian-type functions to the $1/r$ function with the least-squares method [113]. When three Gaussian-type functions are applied, Eq. (20) transforms as follows:

$$V^{el}(r) = \sum_{i}^{N} q_i(\gamma_1 e^{-\alpha_1 (r - R_i)^2} + \gamma_2 e^{-\alpha_2 (r - R_i)^2} + \gamma_3 e^{-\alpha_3 (r - R_i)^2}). \tag{21}$$

Substituting Eq. (21) into the *MEP* function analogue of Eq. (15), the *MEP-SI* can be calculated analytically without application of an arbitrary grid. This method was faster by two orders of magnitude than the corresponding grid-based method and the calculated *MEP-Si$_{(C)}$* values were in very good correlation with each other [113].

Several authors used the Spearman rank correlation coefficient as a *MEP-SI* in similarity investigations [112, 114, 115]:

$$MEP\text{-}SI_{(S)} = 1 - \frac{6\sum_{i}^{M}[r(V_{Ai}) - r(V_{Bi})]^2}{N(N^2 - 1)}. \tag{22}$$

In this algorithm the *MEP* values calculated in the grid points for the two molecules are arranged in increasing numerical order and $r(V_{Ai})$ is the position (rank) of the *EP* value generated by molecule A at point i. The $r(V_{Ai}) - r(V_{Bi})$ value is the difference in ranks corresponding to point i. Since Eq. (22) does not contain concrete numerical values, the *MEP-SI$_{(S)}$* characterizes the similarity of the shapes of the *MEP* functions.

Recently, Petke analyzed the properties of the Carbo- and Hodgkin-type *MEP-SIs* in detail and introduced a new *SI* [116]:

$$MEP\text{-}SI_{(P)} = \frac{\sum_{i}^{M} V_{Ai} V_{Bi}}{max\left(\sum_{i}^{M} V_{Ai}^2, \sum_{i}^{M} V_{Bi}^2\right)}. \tag{23}$$

Figure 12 reveals that when $V_{Ai} = \alpha V_{Bi}(\forall i)$ (α is a constant and $\neq 0$) the Petke *MEP-SI* is an even more sensitive measure of similarity than the Hodgkin index. This is true particularly in the region $\alpha \in [-1, 1]$ where *MEP-SI$_{(P)}$* varies linearly with α. The *SIs* defined by Eqs. (17), (18) and (22) may be called cumulative indices since in each case the *SI* is computed by accumulating products of *MEP* values for a number of grid points [116].

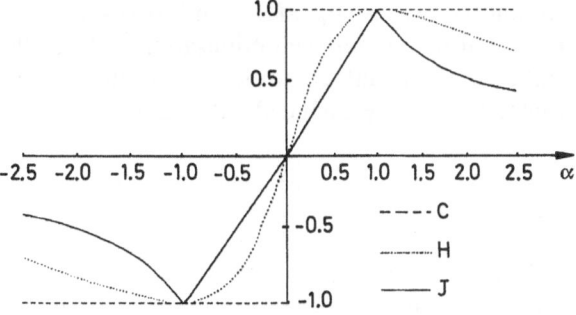

Fig. 12. Variation of similarity indices *MEP-SI$_{(C)}$* (C), *MEP-SI$_{(H)}$* (H), and *MEP-SI$_{(P)}$* (J) with α. (Reproduced from [116]; copyright-John Wiley & Sons)

As an alternative, individual SIs may be calculated for each grid point [116–118]. These are called discrete SIs. Discrete SIs corresponding to cumulative SIs defined by Eqs. (17) and (23) are as follows:

$$d\text{-}MEP\text{-}SI^i_{(H)} = \frac{2 V_{Ai} V_{Bi}}{V^2_{Ai} + V^2_{Bi}} \tag{24}$$

$$d\text{-}MEP\text{-}SI^i_{(P)} = \frac{V_{Ai} V_{Bi}}{\max(V^2_{Ai}, V^2_{Bi})}. \tag{25}$$

The discrete SIs calculated with Eqs. (24) or (25) give a set for molecules A and B. Petke proved that the discrete $MEP\text{-}SI$ defined by Eq. (25) has general linear properties independent of any functional relation between V_{Ai} and V_{Bi} [116].

The set of $d\text{-}MEP\text{-}SI_{(P)}$s, however, represents the mutual similarity of the discrete grid MEP functions. For the guanine molecule Petke showed that, in precise similarity study, one had to analyze statistically the distribution of $d\text{-}MEP\text{-}SI_{(P)}$ values [116]. If a single number is needed to measure the similarity one should use the average value of $d\text{-}MEP\text{-}SI_{(P)}$s or the cumulative Petke $MEP\text{-}SI$ [116].

5 Software Packages for Molecular Similarity Studies

There are several packages applicable to similarity studies [71, 101, 108, 110, 112, 113, 115, 119, 120].

The MOLSIMIL program developed by Carbo and Calabuig [101] is written in FORTRAN-77, and therefore the source code is transferable. The Carbo-type SI is calculated from CNDO-like molecular electron density functions in the program. It is also possible to determine the maximum value of $SI_{(C)}$ with the help of the bisection method.

The ASP (Automated Similarity Package) program [108, 110] is suitable for calculating grid-based (Carbo- and Hodgkin-type) MEP- and $MEF\text{-}SI$s using atomic charges. The simplex method is used for the optimization of SI in the package. It is possible to take into account conformation flexibility. The energetically unreasonable conformations are excluded with the help of Boltzmann weighting factors.

The MEPCOMP [115] and MEPCONF [112] packages developed by Manaut et al. are also suitable for similarity investigations. The MEFCOMP package is able to take into account conformational degrees of freedom as well. For the optimization of the grid-based $MEP\text{-}SI_{(P)}$, a gradient method is applied. For computing MEP several approximations can be used, from the simplified approximation based on atomic charges to the most sophisticated one using Eq. (1).

Acknowlegements. Our work has been sponsored by the National Science Foundation of Hungary (grants: OTKA 616/91 and F4297/92) and by the Pro Renovanda Cultura Hungariae Foundation. The support is gratefully acknowledged.

6 References

1. Johnson MA, Maggiora GM (eds) (1990) Concepts and applications of molecular similarity. John Wiley and Sons, New York
2. Richards WG (1989) Computer-aided molecular design. IBC Technical Services, London
3. Náray-Szabó G (1989) J Mol Graphics 7: 76
4. Dean PM (1987) Molecular foundations of drug-receptor interaction. Cambridge University Press, Cambridge
5. Scrocco E, Tomasi J (1978) Adv Quantum Chem 11: 115
6. Westbrook JD, Levy RM, Krogh-Jespersen K (1992) J Comp Chem 13: 979
7. Weinstein H, Politzer P, Srebrenik S (1975) Theor Chim Acta 38: 159
8. Sen KD, Politzer P (1989) J Chem Phys 90: 4370
9. Pathak RK, Gadre SR (1990) J Chem Phys 93: 1770
10. Silberbach, H (1991) J Chem Phys 94: 8638
11. Shirsat RN, Bapat SV, Gadre SR (1992) Chem Phys Lett 200: 373
12. Scrocco E, Tomasi J (1973) Topics in Current Chemistry 42: 95
13. Politzer P, Daiker KC (1981) In: Deb BM (ed) The force concept in chemistry. Van Nostrand Reinhold Co., New York, p 294
14. Politzer P, Truhlar DG (eds) (1981) Chemical applications of atomic and molecular electrostatic potentials. Plenum Press, New York
15. Poltizer P, Murray JS (1991) In: Lipkowitz KB, Boyd DB (eds) Reviews in Computational Chemistry II. VCH, New York, p 273
16. Johnson BG, Gill PMW, Pople JA (1993) Chem Phys Lett 206: 239
17. Gadre SR, Bapat SV, Sundararajan K, Shrivastava IH (1990) Chem Phys Lett 175: 307
18. Gadre SR, Bapat S, Shrivastava I (1991) Comput Chem 15: 203
19. Frisch MJ, Trucks GW, Head-Gordon M, Gill PMW, Wong MW, Foresman JB, Johnson BG, Schlegel HB, Robb MA, Reploge ES, Gomperts R, Andres JL, Raghavachari K, Binkley JS, Gonzalez C, Martin RL, Fox DJ, DeFrees DJ, Baker J, Stewart JJP, Pople JA (1992) Gaussian 92, Gaussian, Inc., Pittsburg, PA
20. Pople JA, Santry DL, Segal GA (1965) J Chem Phys 43: S129; Pople JA, Segal GA (1965) J Chem Phys 43: S136; Pople JA, Segal GA (1966) J Chem Phys 44: 3289
21. Pople JA, Beveridge DL, Dobosh PA (1967) J Chem Phys 47: 2026
22. Giessner-Prettre C, Pullman A (1972) Theor Chim Acta 25: 83
23. Löwdin PO (1970) J Chem Phys 56: 365
24. Chung-Phillips A (1989) J Comp Chem 10: 17
25. Politzer P (1977) In: Rheingold AL (ed) Homoatomic rings, chains, and macromolecules of main-group elements, Elsevier, Amsterdam, p 95
26. Ridley JE, Zerner MC (1973) Theor Chim Acta 32: 111
27. Culberson JC, Zerner MC (1985) Chem Phys Lett 122: 436
28. Bonaccorsi R, Pullman A, Scrocco E, Tomasi J (1972) Theor Chim Acta 24: 51
29. Benoit RL, Fréchette M (1986) Can J Chem 64: 2348
30. Del Bene JE (1983) J Phys Chem 87: 367
31. Ford GP, Wang B (1993) J Comp Chem 14: 1101
32. Dewar MJS, Thiel W (1977) J Am Chem Soc 99: 4899
33. Dewar MJS, Zoebisch EG, Healy EF, Stewart JJP (1985) J Am Chem Soc 107: 3902
34. Stewart JJP (1989) J Comp Chem 10: 209, 221
35. Pople JA, Beveridge DL (1970) Approximate molecular orbital theory, McGraw-Hill, New York

36. Hehre WJ, Radom L, Schleyer PR, Pople JA (1986) Ab initio molecular orbital theory, Wiley, New York
37. Luque FJ, Illas F, Orozco M (1990) J Comp Chem 11: 416
38. Luque FJ, Orozco M (1990) Chem Phys Lett 168: 269
39. Alemán C, Luque FJ, Orozco M (1993) J Comp Chem 14: 799
40. Ferenczy GG, Reynolds CA, Richards WG (1990) J Comp Chem 11: 159
41. Reynolds CA, Ferenczy GG, Richards WG (1992) J Mol Struct (Theochem) 256: 249
42. Tasi G, Kiricsi I, Förster H (1991) Magy Kém Foly 97: 441
43. Tasi G, Kiricsi I, Förster H (1992) J Comp Chem 13: 371
44. Tasi G, Pálinkó I, Nyerges L, Fejes P, Förster H (1993) J Chem Inf Comput Sci 33: 296
45. Bingham RC, Dewar MJS, Lo H (1975) J Am Chem Soc 97: 1285
46. Tasi G, Pálinkó I, Halász J Náray-Szabó G (1992) Semiempirical quantum chemical calculations on microcomputers. CheMicro Ltd, Budapest
47. Nelder JA, Mead R (1965) Comput J 7: 308
48. Cummins PL, Gready JE (1990) Chem Phys Lett 174: 355
49. Parr RG, Yang W (1989) Density functional theory of atoms and molecules. Oxford University Press, New York
50. Murray JS, Seminario JM, Concha MC, Politzer P (1992) Int J Qunatum Chem 44: 113
51. Fliszár S (1983) Charge distributions and chemical effects. Springer-Verlag, Berlin
52. Bowen JP, Allinger NL (1991) In: Lipkowitz KB, Boyd DB (eds) Reviews in Computational Chemistry II. VCH, New York, p 81
53. McCammon JA, Harvey SC (1987) Dynamics of proteins and nucleic acids. Cambridge University Press, Cambridge
54. Montagnani R, Tomasi J (1993) J Mol Struct (Theochem) 279: 131; Kozaki T, Morihashi K, Kikuchi O (1988) J Mol Struct 168: 265
55. Hall GG (1985) Adv Atomic Mol Phys 20: 41
56. Williams DE, Yan Y-M (1988) Adv Atomic Mol Phys 23: 87
57. Williams DE (1991) In: Lipkowitz KB, Boyd DB (eds) Reviews in Computational Chemistry II. VCH, New York, p 219
58. Momany FA (1978) J Phys Chem 83: 592
59. Cox SR, Williams DE (1981) J Comp Chem 2: 304
60. Singh UC, Kollman PA (1984) J Comp Chem 5: 129
61. Chirlian LE, Francl MM (1987) J Comp Chem 8: 894
62. Woods RJ, Khalil M, Pell W, Moffat SH, Smith VH, Jr. (1990) J Comp Chem 11: 297
63. Besler BH, Merz, KM, Jr., Kollman PA (1990) J Comp Chem 11: 431
64. Orozco M, Luque FJ (1990) J Comp Chem 11: 909
65. Chipot C, Maigret B, Rivail J-L, Scheraga HA (1992) J Phys Chem 96: 10276
66. Merz KM, Jr. (1992) J Comp Chem 13: 749
67. Bertsekas DP (1982) Constrained optimization and Lagrange multiplier methods. Academic Press, New York
68. Williams DE, Weller RR (1983) J Am Chem Soc 105: 4143
69. Breneman CM, Wiberg KB (1990) J Comp Chem 11: 361
70. Connolly M (1982) QCPE Program 429
71. Richard AM (1991) J Comp Chem 12: 959
72. Lee J-G. Friesner RA (1993) J Phys Chem 97: 3515
73. Orozco M, Luque FJ (1990) J Comput-Aided Mol Des 4: 441
74. Náray-Szabó G, Ferenczy GG (1992) J Mol Struct (Theochem) 261: 55; Ferenczy GG, Rivail J-L, Surján P, Náray-Szabó G (1992) J Comp Chem 13: 830
75. Su Z (1993) J Comp Chem 14: 1036
76. Coppens P (1992) Annu Rev Phys Chem 43: 663
77. RATTLER, Oxford Molecular Ltd, The Magdalen Centre, Oxford Science Park, Sandford-on-Thames, Oxford OX4 4GA, United Kingdom
78. Merz KM, Besler BH (1990) MOPAC-ESP, QCPE program 589
79. Mulliken RS (1955) J Chem Phys 23: 1833
80. Szabó A, Ostlung NS (1982) Modern quantum chemistry. Macmillan, New York
81. Huzinaga S, Sakai Y, Miyoshi E, Narita S (1990) J Chem Phys 93: 3319
82. Surján PR (1989) Second quantized approach to quantum chemistry. Springer-Verlag, Berlin
83. Mayer I (1983) Chem Phys Lett 97: 270
84. Rauhut G, Clark T (1993) J Comp Chem 14: 503

85. Foster JP, Weinhold F (1980) J Am Chem Soc 102: 7211
86. Ferenczy GG (1991) J Comp Chem 12: 913
87. Stone AJ (1981) Chem Phys Lett 83: 233; Stone AJ, Price SL (1988) J Phys Chem 92: 3325
88. Chipot C, Ángyán JG, Ferenczy GG, Scheraga HA (1993) J Phys Chem 97: 6628
89. Sawaryn A, Sokalski WA (1989) Comp Phys Comm 52: 397
90. Sokalski WA, Shibata M, Ornstein RL, Rein R (1993) Theor Chim Acta 85: 209
91. Sokalski WA, Hariharan PC, Kaufmann JJ (1987) Int J Quantum Chem, Quantum Biol Symp 14: 111
92. Sokalski WA, Maruszewski K, Hariharan PC, Kaufmann JJ (1989) Int J Quantum Chem, Quantum Biol Symp 16: 119
93. Sokalski WA, Poirier RA (1983) Chem Phys Lett 98: 86
94. Sokalski WA, Sawaryn A (1987) J Chem Phys 87: 526
95. Vigne-Maeder F, Claverie P (1988) J Chem Phys 88: 4934
96. Mezei M, Campbell ES (1977) Theor Chim Acta 43: 227
97. Spackman MA (1986) J Chem Phys 85: 6587
98. Sokalski WA, Sneddon SF (1991) J Mol Graphics 9: 74
99. Carbo R, Leyda L, Arnau M (1980) Int J Quantum Chem 17: 1185
100. Carbo R, Domingo L (1987) Int J Quantum Chem 32: 517
101. Carbo R, Calabuig B (1989) Comp Phys Comm 55: 117
102. Bowen-Jenkins PE, Cooper DL, Richards WG (1985) J Phys Chem 89: 2195
103. Hodgkin EE, Richards WG (1986) J Chem Soc, Chem Commun 1342
104. Hodgkin EE, Richards WG (1987) Int J Quantum Chem, Quantum Biol Symp 14: 105
105. Stewart JJP (1983) QCPE Bull 3: 43
106. Dewar MJS, McKee ML, Rzepa HS (1978) J Am Chem Soc 100: 3607
107. Dewar MJS, Reynolds CH (1986) J Comp Chem 7: 140
108. Burt C, Richards WG, Huxley P (1990) J Comp Chem 11: 1139
109. Burkert U, Allinger NL (1982) Molecular Mechanics. Am Chem Society, Washington, DC
110. Burt C, Richards WG (1990) J Comput-Aided Mol Des 4: 231
111. Reynolds CA, Essex JW, Richards WG (1992) J Am Chem Soc 114: 9075; Colonna F, Evleth EM (1993) Chem Phys Lett 212: 665
112. Sanz F, Manaut F, Sanchez JA, Lozoya E (1991) J Mol Struct. (Theochem) 230: 437
113. Good AC, Hodgkin EE, Richards WG (1992) J Chem Inf Comput Sci 32: 188
114. Namasivayam S, Dean PM (1986) J Mol Graphics 4: 46; Dean PM, Callow P, Chau P-L (1988) J Mol Graphics 6: 28
115. Manaut F, Sanz F, José J, Milesi M (1991) J Comput-Aided Mol Des 5: 371
116. Petke JD (1993) J Comp Chem 14: 928
117. Reynolds CA, Burt C, Richards WG (1992) Quant Struct-Activ Relat 11: 34
118. Good AC (1992) J Mol Graphics 10: 144
119. ASP (Automated Similarity Package), Oxford Molecular Ltd, The Magdalen Centre, Oxford Science, Park, Sandford-on-Thames, Oxford OX4 4GA, United Kingdom
120. Mezey PG (1986) Int J Quantum Chem, Quantum Biol Symp 12: 113

The Use of Graph Theoretical Methods for the Comparison of the Structures of Biological Macromolecules

Peter J. Artymiuk[1]* Andrew R. Poirrette[2] David W. Rice[1] and Peter Willett[2]

[1] Krebs Institute for Biomolecular Research, Department of Molecular Biology and Biotechnology, University of Sheffield, Sheffield S10 2TN, United Kingdom
[2] Krebs Institute for Biomolecular Research, Department of Information Studies, University of Sheffield, Sheffield S10 2TN, United Kingdom

Table of Contents

List of Symbols and Abbreviations 74

1 Introduction 75

2 Biological Macromolecules 75
  2.1 Macromolecular Structure: Levels of Description 76
    2.1.1 Primary Structure 76
    2.1.2 Secondary Structure 77
    2.1.3 Tertiary Structure 77

3 Protein Structure Databases 77
  3.1 Searching Methods for One-Dimensional Sequences 78
  3.2 Three Dimensional Databases 79
    3.2.1 Determination of Structures of Biological
        Macromolecules 79
    3.2.2 The Protein Data Bank 80

4 Comparison of Protein Folds 81
  4.1 Methods for Comparing Protein Folds in Three Dimensions . . 82
  4.2 Use of Graph Theoretic Methods for Detection
    of Similarity in Protein Folds 84

---

* To whom correspondence should be addressed

Topics in Current Chemistry, Vol. 174
© Springer-Verlag Berlin Heidelberg 1995

4.2.1 Use of PROTEP: Structural Similarities Involving
HIV Reverse Transcriptase 87

5 Investigation of Local Protein Structure 89
5.1 Comparison of Proteins at the Sidechain Level using
Algorithms Derived from Graph Theory 91
5.1.1 Pseudo-Atom Representation of Sidechains 91
5.1.2 The ASSAM Search Program 92
5.1.3 Structural Similarity Between Binding Sites
in Influenza Sialidase and Isocitrate Dehydrogenase . . . 96
5.1.4 Future Directions 98
5.2 Other Applications 98

6 Conclusions 100

7 References 101

List of Symbols and Abbreviations

3-D   three-dimensional
AIDS   acquired immune deficiency syndrome
ASCII   American Standard Code for Information Interchange
ATP   adenosine triphosphate
CPU   central processor unit
CSD   Cambridge Structural Database
DNA   2′-deoxy-ribonucleic acid
HIV   Human Immunodeficiency Virus
ICDH   isocitrate dehydrogenase
NMR   nulcear magnetic resonance spectroscopy
PDB   Protein Data Bank
RDBMS   relational database management system
RNA   ribonucleic acid
RNAase   ribonuclease
RT   reverse transcriptase
SSE   secondary structure element
UMP   uracil monophosphate

The use of graph-theoretical algorithms in the similarity searching of databases of 3-dimensional structures of macromolecules is discussed. The emphasis is on the structures of protein molecules, for which the vast majority of 3-dimensional information is available. An initial survey is made of the types of information available on macromolecular structure, and of the methods conventionally used to compare them. Next the use of graph theoretical algorithms to identify previously unrecognized similarities between the overall folds of proteins is described, using the structure of the HIV reverse transcriptase enzyme as an example. The further application of these methods to the problem of searching for three-dimensional patterns of sidechain functional groups is then considered, using the active site of influenza sialidase as an example.

1 Introduction

The principal aims of research in molecular biology are to explain at the molecular level how a cell encodes and organises its genetic information in genes, how this information is passed onto following generations, how genes are expressed as cellular proteins, and what molecular mechanisms enable these proteins to carry out their biological roles. The elucidation of the mechanisms by which organisms function both at the cellular and molecular level is of immense medical, biological and commercial importance, with the result that molecular biology is one of the most rapidly growing areas of modern science. The complexity and quantity of the information concerning biological macromolecules is already very considerable, and this means that the application of computational database technology has necessarily begun to make a major impact in the field.

The last five to ten years have seen rapid developments in database techniques for examining and comparing macromolecular structures. The ability to make such comparisons is of great value in molecular biology, biochemistry and biotechnology. Early work in this area concentrated on the detection of similarities in the one-dimensional sequences of biological macromolecules, but three-dimensional methods have now become important with the recent rapid increases in the number of known three-dimensional macromolecular structures, specifically of proteins. Some three-dimensional structure comparison methods have built on the earlier one-dimensional ones, and these are briefly described below. However the main topic discussed will be the recent development of graph-theoretical methods for the detection of molecular similarity in protein molecules. These methods are a direct development of long-established techniques for the representation and searching of small molecule structures [1].

2 Biological Macromolecules

There are three classes of macromolecule used in cellular metabolism: nucleic acids, proteins and carbohydrates. All three, identified below, are covalently linked polymeric assemblages of small "building blocks".

Nucleic acids consists of a linear backbone comprising a succession of 5-carbon sugar rings alternating with phosphate groups. Attached to each sugar is a "base" which is a derivative of a heterocyclic aromatic pyrimidine. The most important function of nucleic acids is to carry genetic information: the linear sequence of based makes up a "gene" which contains the information necessary to synthesize either another nucleic acid molecule or a particular protein.

A protein consists of a linear chain of amino acids, each of which consists of a central alpha-carbon atom linked to an amino group, a carboxyl group and one of twenty possible side chain groups. Proteins (which include enzymes) are the most varied and versatile of the cell's components and perform a multitude of biological roles which include enzyme catalysis, the control of cell shape and motion, cellular communication, immune defence and cell regulation.

Carbohydrates are made up of monosaccharide building blocks. Unlike proteins and nucleic acids, which form consistently-linked linear polymers, carbohydrates often form branched and variably interlinked chains. In addition to their traditionally recognised functions as structural and as energy storage molecules, carbohydrates are now known to perform a multitude of essential roles in biology, often by forming glycoproteins in which carbohydrate chains are covalently linked to protein molecules. These functions include communication and recognition within and between cells, tissue and cell differentiation, involvement in the immune response.

We shall concentrate almost exclusively on the 3-D structure of proteins. Because of their wide range of functions they are the class of macromolecule which are of the greatest medical and technological interest, as they form the targets for drug and inhibitor design. Because of this they are also the type of macromolecule for which by far the greatest amount of three-dimensional information is available. However similarity searching of all classes of biological macromolecules can be expected to be confronted by similar problems, and it is reasonable to expect that the methods developed for the analysis of 3-D protein structures will also be applicable, with modifications, to the 3-D structures of nucleic acids and complex carbohydrates.

2.1 Macromolecular Structure: Levels of Description

The three classes of macromolecular structures can be considered at a number of increasingly complex levels, described as their primary, secondary and tertiary structures.

2.1.1 Primary Structure

Amino acids link together linearly to form proteins, nucleotides link linearly to form RNA and DNA, and sugars link in a more complicated way to form complex carbohydrates. The specific sequence in which these units link together to form the final polymeric macromolecule is called its *primary structure*. In a way that is still very ill-understood, the primary structure ultimately controls the macromolecule's three-dimensional structure and thereby largely determines its function. There is therefore great interest in analyzing primary structural information in order to detect similarities and relationships between macromolecules. However, as we shall see later, although similar primary structures imply similar three-dimensional structures, it is possible for three-dimensional structures to resemble each other without any sequence similarity.

2.1.2 Secondary Structure

Much of the backbone of a protein or a nucleic acid can be divided into blocks of regular backbone structure. In proteins there are three main types of structure: (i) the alpha helix, which is a tight helix (one turn every 3.6 residues) stabilized by hydrogen bonds between mainchain carbonyl groups and the mainchain amino group of the sidechain four residues further on in the sequence; (ii) the beta sheet in which lengths of polypeptide chain ("beta strands") lie next to one another, in either parallel or anti-parallel arrangement, with hydrogen bonds between mainchain NH and CO groups on adjacent strands; (iii) finally there are turns, loops and "random coil" structures in which the hydrogen bond pattern is less repetitive and the structures are thus less regular. In nucleic acids it is possible to identify mutually complementary stretches of sequence which give rise to double helical structures through Watson-Crick base-pairing [2]. Not enough is yet known of 3-D carbohydrate structure to know if a description at the level of secondary structure would be either valid or useful.

It is important to note that the secondary structural description of a macromolecule involves the use of limited and localised three-dimensional information to give a schematic indication of the organisation of a macromolecular structure. However such a secondary structural description says nothing concerning the relative 3-D orientation of these secondary structural elements.

2.1.3 Tertiary Structure

The tertiary structure of a biological macromolecule is its actual three-dimensional structure: this includes not only information about the relative orientations of secondary structure elements in their backbone, but also, for example, detailed information concerning the orientations and interactions of sidechains, bases or sugar moieties. An understanding of the tertiary structure of a macromolecule is an essential step in comprehending its function, and its modes of binding to ligands or other macromolecules.

Some biological assemblies consist of an array of a number of identical or non-identical polypeptide or nucleotide chains. The way in which these subunits are oriented with respect to one another is defined as the "quaternary structure". This fourth level of description is not described further here.

3 Protein Structure Databases

Because of their size and complexity, the analysis of the sequences and structures of proteins, nucleic acids and carbohydrates represent a major informational challenge. They are, however, ideally suited to the use of database techniques, because of the manner in which their molecular structures are built up from a small number of basic chemical building blocks.

Because the (one-dimensional) sequence of a biological macromolecule determines its ultimate 3-D structure, the detection of sequence similarity implies a 3-D resemblance as well as an evolutionary relationship, and it is therefore relevant to review briefly methods for sequence comparison. However, as we shall see below, the converse is not the case and it is possible for proteins to have very similar 3-D structures without there being any sequence similarity. It is the detection of these more distant structural relationships that form our main subject.

3.1 Searching Methods for One-Dimensional Sequences

The most basic structural information that can be obtained concerning a protein, nucleic acid or carbohydrate molecule is the sequence of the basic building blocks (amino acids, nucleotides or sugars, respectively) that comprise its covalent structure. Sequence data for biological macromolecules are much more easily determined than are their 3-D structures. In the case of proteins this information can be obtained by the stepwise chemical removal and identification of the individual amino acids from the N-terminal end of the complete protein chain or of fragments derived from it [3]. However, nucleic acid sequencing is now the major means of sequencing proteins, as the genetic message in the DNA sequence can easily be translated into the equivalent protein sequence using the genetic code [4]. The sequence data for nucleic acids can be derived either from the enzymatic method of rapid sequencing developed by Sanger in 1977 [5] or the chemical method developed by Maxam and Gilbert [6] in 1980. The analysis of a carbohydrate sequence is much more difficult, requiring the identification of the individual sugar residues (most often done by methylation and hydrolysis, followed by analysis using mass spectrometry and gas-liquid chromatography). Identification of the sequence and anomeric configuration of the monosaccharides can be achieved by the use of specific glycosidase enzymes or by 2-dimensional nuclear magnetic resonance spectroscopy [7].

The advances in protein, and especially DNA, sequencing technology means that there is now a vast amount of primary structural information relating to biological macromolecules and it is hence essential for laboratories in the field to make use of computers to analyse data on protein and nucleic acid sequences. At present (June 1994) there are more than 80 000 sequences in the OWL protein sequence database [8] and there are more than 170 000 nucleic acid sequences in the EMBL (European Molecular Biology Laboratory) database [9].

The methods of nucleic acid sequencing have now advanced to such an extent that the United States and other national governments are funding a 10–20 year project to sequence the entire human genome [10, 11] at an estimated cost of $ 3 billion. The genome is the complete genetic blueprint of a human being. It consists of twenty four distinct chromosomes (twenty two pairs of autosomes plus the two sex chromosomes X and Y), each chromosome consisting of one double stranded DNA molecule associated with various protein molecules. The total size of the genome is around 3 billion bases which compose about 100 000 genes, each gene specifying a distinct protein or nucleic acid molecule. The Human Genome Project aims to locate the positions of all the genes in each chromosome (including erroneous genes responsible for genetic diseases), and to obtain a complete readout of the base sequence of each chromosome. It seems probable that the medical benefits of the program will be immense, but it is difficult to assess the possible ethical social and political consequences of the use and abuse of the genetic information it will reveal.

Numerous programs have been written by many workers in the field in order to analyse different aspects of protein and nucleic acid sequences. Lesk [12] reviews the many available methods, several of which address the problem of molecular similarity at the primary structural level. In particular, when a new protein sequence has been determined, it is often very informative to search the databases of known sequences in order to detect any unsuspected similarities that may give insights into function or into evolutionary relationships: the methods used for this purpose include simple character string comparisons, the use of comparison matrices which compare the characteristics of different amino acids [13], and the use of dynamic programming [14]. The more detailed comparison of a pair of similar sequences, often done using 'diagonal plots' [15] can yield further valuable information. Detailed comparisons between a pair of protein sequences (as opposed to sequence searching, which involves the less detailed comparison of one sequence with a complete database of others) can be carried out by a variety of methods including the method of Maizel and Lenk [16] or the rapid algorithm of Sussmann and Unger [17] to produce diagonal dot plots. Optimal alignments of sequences can be determined by the methods of Needleman and Wunsch [14] or the local homology method of Smith and Waterman [18]. Sequence database searching is also supported, primarily using the method of Wilbur and Lipman [19, 20].

3.2 Three-Dimensional Databases

3.2.1 Determination of Structures of Biological Macromolecules

The determination of the three-dimensional structures of biological macromolecules is a major area of scientific endeavour. Several hundred different structures have now been determined, principally of proteins though also of DNA, RNA and carbohydrates. The molecules concerned are often extremely large: the largest asymmetric protein structure so far determined is the photoreaction centre [21] which consists of over 10 000 non-hydrogen atoms. Symmetrical virus structures of 100 000 non-hydrogen atoms have also been determined [22]. The chief method of structure elucidation is single crystal X-ray diffraction [23, 24]. However, Nuclear Magnetic Resonance spectroscopy (NMR) is becoming a valuable method for the structure determination of smaller proteins [25]: at present NMR is effectively limited to molecular structures of up to 1000 non-hydrogen atoms, but rapidly developing techniques offer the possibility of studying somewhat larger molecules in the future. The end point of all such structural studies is the production of coordinate sets for each of the atoms in the molecule, and therefore large amounts of storage are required, and powerful methods must be developed to examine such systems. The technique of electron microscopy coupled with electron diffraction is also becoming an important method for the determination of the structures of membrane proteins [26, 27].

3.2.2 The Protein Data Bank

The Protein Data Bank (PDB) was established at the Brookhaven National Laboratory in 1971 [28, 29]. It functions as the internationally recognized archive of the three-dimensional structures of biological macromolecules. The great majority of the depositions in the PDB originate from X-ray single crystal diffraction studies. Although many of these give the protein structures in atomic detail, a proportion of entries are incomplete in the sense that only an alpha carbon trace of the molecular fold is deposited, and no data are therefore given on sidechain positions. Recently, as observed above, NMR structures are also becoming a significant source of smaller macromolecular structures [25]. As well as these experimentally determined coordinates, which comprise the greater part of the PDB, a few hypothetical, model-built structures are also included.

Most of the coordinate sets in the PDB are those of proteins. However, the advent of simple methods of oligonucleotide synthesis has considerably facilitated the preparation of pure samples of nucleic acids. The number of single crystal structure determinations of nucleic acids and of protein-nucleic acid complexes is therefore increasing. Whilst protein structures have long been accepted as the sole province of the PDB, the nucleic acid, and also the oligosaccharide, depositions involve a small amount of overlap with depositions in the Cambridge Structural Database (CSD) [30]. However, one advantage of PDB depositions is that, unlike the CSD, all coordinates in the PDB are transformed to conventional orthornormal coordinate frames which facilitates their use by non-crystallographers.

The size of the PDB is increasing very rapidly: in the January 1994 "full release" of the databank there were 2327 coordinate entries of which 605 had appeared since the previous, October 1993, release. The entries comprise 2143 protein structures, together with 156 DNA, 18 RNA and 10 carbohydrate structures. The PDB also contains bibliographic references to more than 100 macromolecular structures which have been published, but which have not been deposited in the PDB by the authors. Recently there has been strong pressure on journals to make deposition of coordinates in the PDB a condition of publication [31].

It is important to note that many of the depositions in the PDB are not unique: for any protein there may be a number of coordinate sets representing different stages of crystallographic refinement, the binding of different substrates to the protein, or different site-directed mutants. Also, in certain cases, different species variants of the same protein may differ in only a few amino acids. It is difficult to quantify exactly, but at present there are probably less than 300 distinct protein structures in the PDB.

It is also essential to be aware of the limitations of the entries in the PDB. The PDB attempts to ensure that coordinate depositions are internally consistent by carrying out a limited checking procedure (e.g. the amino acid sequence records should agree with the coordinates). However, unlike the CSD, it is not their policy at present to monitor the quality of the structures which are supplied by the depositors. Errors can arise from a number of sources. First, the initial report of a protein structure is usually based on the interpretation of a medium resolution (3.0–2.5 Å) electron density map calculated with phases computed using the method of multiple isomorphous replacement. This initial map is often of poor quality in some regions and thus errors in interpretation can occur. These can later be corrected by structure factor least squares crystallographic refinement and extension of the resolution of the data [32]. Second, proteins are inherently somewhat flexible molecules: this means that even in a good electron density map certain parts of the molecule may be either dynamically or statically ordered: exposed loops and turns are especially prone to this, but on occasion whole domains of several hundred residues may be invisible in the map [33]. Another consequence of the flexibility of protein molecules is that the vast majority of protein crystals diffract much less than typical small molecule crystals: many do not diffract to spacings beyond 2.0 Å, and diffraction to spacings beyond 1.0 Å is only seen in a few exotic cases. There are therefore relatively low observation/parameter ratios in protein refinements. The result is that even the best refined protein coordinate sets do not match typical small molecule structures in precision. It is consequently very unusual to have data on hydrogen atom coordinates (unless a neutron diffraction study has been conducted) or on anisotropic temperature factors. The user of the PDB should therefore be wary and realize that there are a few coordinate sets in the PDB with exceedingly improbable conformational torsional angles and impossible inter-atomic contacts. However, over the last few years it has become the normal practice to conduct restrained structure factor least squares refinement on protein structures and consequently the overall quality of coordinate sets has increased considerably.

4 Comparisons of Protein Folds

The PDB is not itself a database: it simply consists of more than 2000 ASCII files each of which contains the coordinates of a macromolecule. At present the PDB does not distribute any mechanisms for conducting searches for specific features within the databank. However in recent years there has been very intense activity amongst researchers in the field of biological structure to derive methodologies to systematize and to enhance understanding of how the amino acid sequence of a protein governs the final 3-D structure, and therefore the function of that protein remains the greatest unsolved problem in structural molecular biology [34].

It is possible at present to identify two main levels at which molecular similarity is of importance in proteins. First, detection of large-scale similarities between different protein structures, i.e.: similarities in the way that the linear polypeptide sequence is folded up to form a three-dimensional structure. This is the subject of the remainder of Sect. 4. Second, comparative analysis of local aspects of protein structure, for example the examination of specific binding sites, or of the environments of particular sidechains. These methods are described in Sect. 5.

4.1 Methods for Comparing Protein Folds in Thee Dimensions

The fold of a protein is the way in which the regions of helix, strand and random-coil structure within its polypeptide chain are arranged in three dimensions to form its tertiary structure (see Sect. 2.1.3). This is the simplest, and yet often a very revealing, level at which the three-dimensional structures of different proteins can be compared with one another: as is indicated below, such similarities may be indicators of remote evolutionary relationships, give clues to functional analogies, or insights into the processes of protein folding.

The comparison of protein folds has proved to be difficult: the three-dimensional structures are frequently complicated, and quite significant differences can exist between structures that are, on the basis of sequence similarity, clearly related in evolutionary terms. On the other hand structures may sometimes resemble each other very closely, but fail to display any sequence similarity: the classic example of this is the "parallel beta barrel" structure which has now been found in more than twenty proteins with no amino-acid sequence homology [35]. In these cases the interpretation of the meaning of a similarity can be less than straightforward: it may indicate that the proteins are evolutionarily related ("divergent evolution"), that they are unrelated but have evolved similar structures because they carry out similar functions ("convergent evolution") or the common structure may simply be a particularly stable one that is adopted by a large number of proteins. In addition to three similarities between complete protein folds, there may also be partial similarities.

Previous work in the field of similarity searching in proteins has led to the development of techniques to carry out numerous comparative tasks in molecular biology, ranging from the creation of sequence alignments to the comparison of 3-D protein structures in geometric terms. In the latter area, much of the work to date has concentrated on the detection of similarity in folding patterns. The main emphasis has been on the comparison of protein folds by alignment of large portions of protein structures to locate maximal lengths of superimposable main chain. Rao and Rossmann [36] noted that similar three-dimensional arrangements of alpha helices and beta strands could occur in different protein structures. They called these arrangements "super-secondary structures", and they are also now known as folding "motifs". The earliest quantitative methods for protein structure comparisons were developed by Rossmann and coworkers [37–39] who developed a complex but powerful procedure for optimally overlapping a pair of protein structures. Remington and Matthews [40, 41] developed a simpler method which superimposes all possible segments of L residues from one protein on all possible segments of L residues of another, where L is a probe length defined by the user. However this latter method was less effective than Rossmann's at allowing for major insertions and deletions in the polypeptide chain between the proteins that were being compared. Both methods require considerable computing time to compare a pair of proteins and are consequently unsuitable for conducting rapid searches for structural similarities between a protein or a motif and all the other proteins in the PDB.

The rapid expansion in the number of known protein structures during the 1980s led to a resurgence of interest in this area, and there has been intense research activity in this field over the last five years. Thus, more recent approaches to the problem of comparing folds have been reported by Lesk [42, 43]. Taylor and Orengo [44] have extended the dynamic programming techniques of Needleman and Wunsch [14]; Sali and Blundell combined dynamic programming with simulated annealing [45], and Vriend and Sander [46] and Alexandrov et al. [47] have both developed methods by which proteins may be compared by clustering together similar substructures; Nussinov and coworkers have used "geometric hashing", a computer-vision technique, to compare protein structures [48, 49], and very recently May and Johnson [50] have described the use of genetic algorithms for aligning macromolecular structures. At the same time as these developments, comparison methods for proteins, based on graph theory, were developed in our laboratories [51–53] and elsewhere [54], and these are discussed in more detail in the following subsections. There has also been interest in the development of protein-based similarity searching techniques in rational drug design, where the main focus of interest has been the development of techniques for docking ligands into receptor sites [55, 56].

The majority of the above methods use protein alpha-carbon coordinates for their searches. However, a number of workers have simplified the problem by representing alpha helices and beta strands as straight line segments. A protein can then be represented by a matrix containing information on the relative positions and orientations of pairs of these linear segments. Comparisons between proteins can be affected by comparing these matrices. Richards and Kundrot [57] describe a program for representing proteins in this manner, using up to six parameters to describe the relationship between each pair of helices and/or strands. They then describe a procedure for searching for arbitrary substructures using such distance matrices. The present authors have adopted a similar simplified approach in some of their graph-theoretical approaches to the problem of comparing protein folds, and this work is described in the next section.

4.2 Use of Graph Theoretical Methods for Detection of Similarity in Protein Folds

For several years we have been involved in a wide-ranging project to develop methods for the representation and searching of the three-dimensional (3-D) protein structures in the Brookhaven Protein Data Bank [28, 29]. Our work derives from the graph-theoretic methods that are used for the storage and retrieval of information pertaining to both two-dimensional (2-D) and 3-D small molecules [1, 58].

A molecule in a chemical information system is represented by a labelled graph, in which the nodes and edges represent the atoms and bonds, respectively, of a 2-D molecule, i.e., a planar chemical structure diagram, or the atoms and inter-atomic distances, respectively, of a 3-D molecule. This graph-theoretic representation enables searching operations on databases of chemical structures to be implemented using isomorphism algorithms, which compare one graph with another to determine the structural relationships that exist between them. Thus subgraph isomorphism algorithms are used in substructure searching systems, which retrieve all molecules from a database that contain a user-defined partial structure, e.g., all molecules containing a quinazoline ring system. Maximal common subgraph isomorphism algorithms are used in reaction-indexing systems, where they are used to identify those part of the reactant and product molecules that remain unchanged during a reaction [59], and in similarity searching systems, which permit a user to identify those molecules in a database that are structurally most similar to an input target molecule [60]. Our work has involved the application of both of these types of algorithm to the searching of 3-D protein structures [52, 53, 61–66].

The graph representations of protein folds are constructed so that the nodes of the graph are the secondary structure elements (i.e., the alpha helices and beta strands) of each protein, whilst the edges are the distances and angles between them. In order to accomplish this, regions of helix and strand in proteins in the Protein Data Bank [28, 29] are assigned using the algorithm of Kabsch and Sander [67]. The position and direction of each secondary structure element (SSE) is then approximated by a vector in 3-dimensional space which corresponds to the axis of an idealized helix or strand superposed on the real helix or strand by least squares. The torsional angles, closest approach distances and distances between midpoints of each pair of SSEs within each protein in the PDB are stored in a database as a labelled graph. The nodes of the graph are the linear representations of the SSEs, and the edges of the graph the distances and angles between them [52].

The two search algorithms that process the protein-structure representation described in the previous paragraph derive from work carried out previously in

Sheffield on 3-D searching in databases of small 3-D molecules [58, 68, 69]. These studies involved a detailed comparison of a range of different subgraph and maximal common subgraph isomorphism algorithms and concluded that the algorithms due to Ullmann [70] and to Bron and Kerbosch [71] were the most efficient of those that were tested. Both of them have accordingly been implemented in a range of software packages (both in-house and commercial) for small-molecule database searching and we have modified them so that they can be used for searching 3-D protein structures. Two programs, POSSUM and PROTEP (which together form the Tripos Associates PROTEP program package) implement the Ullmann and the Bron and Kerbosch algorithms respectively. In Sect. 5.1 we describe the ASSAM program which is used for more detailed sidechain-level searching: currently this program implements just the Ullmann algorithm, although the application of the Bron-Kerbosch algorithm to the processing of sidechain data is under investigation in our laboratories at present.

The Ullmann subgraph-isomorphism algorithm [70] operates by means of a backtracking tree search in which nodes from a database protein structure are tentatively assigned to nodes from the query pattern and the match extended in a depth-first manner until a complete match is obtained or until a mismatch is detected; in this case, the search then backtracks to the previous assignment. Such an approach is common to many graph-matching procedures: the great efficiency of the Ullmann algorithm results from the use of a refinement procedure, which limits the number of levels of the search tree that have to be investigated before a mismatch is identified.

The refinement procedure utilises the fact that if some query node $Q(X)$ has another node $Q(W)$ at some specific distance(s) (and/or angle), and if some database node $D(Z)$ matches with $Q(W)$, then there must also be some node $D(Y)$ at the appropriate distance(s) from $D(Z)$ which matches with $Q(X)$: this is a necessary, but not sufficient, condition for a subgraph isomorphism to be present (except in the limiting case of all the query nodes having been matched, when the condition is both necessary and sufficient). The refinement procedure is called before each possible assignment of a database node to a query node; and the matched substructure is increased by one node if, and only if, the condition holds for all nodes W, X, Y and Z. The basic algorithm terminates once a match has been detected or until a mismatch has been confirmed [70]; it is easy to extend the algorithm to enable the detection of all matches between a query pattern and a database structure, as is required for applications such as those discussed here.

The maximal common subgraph procedure that we have used is based on the clique-detection algorithm of Bron and Kerbosch [71], where a clique is a subgraph of a graph in which every node is connected to every other node and which is not contained in any larger subgraph with this property. The input to the clique-detection procedure is a "correspondence graph". Given a pair of graphs A and B, the correspondence graph, C, is formed in two stages. In the first stage, the set of all pairs of nodes is created, one from A and one from B, such that the nodes of each pair are of the same type. C is then the graph whose nodes are the pairs from the first step. Two nodes (A(I), B(X)), (A(J), B(Y)) are marked as being connected in C if the value of the edges from A(I) to A(J) and B(X) to B(Y) are the same. Maximal common subgraphs then correspond to the cliques in the correspondence graph [72], so that the identification of the maximal common substructure for a pair of 3-D molecules is equivalent to the identification of the largest clique in the correspondence graph linking the two molecules. This largest clique then represents the largest possible overlap of one structure on the other, subject to any geometric tolerances that have been defined by the user.

The first program to be developed using these methods was called POSSUM (Protein Online Substructure Search, Ullmann Method) [52]), in which graph representations of proteins were searched using a modification of the subgraph isomorphism algorithm of Ullmann [70]. POSSUM has been extensively tested [52] against a variety of known motifs including the TIM β-barrel, the nucleotide binding domain, and the various beta-sheet patterns described by Richardson [73]. Using POSSUM we were, for example, able to detect a previously unrecognized and intriguing structural homology between the CheY bacterial signal transduction protein and EFTu, an elongation factor related to G proteins [61]. Moreover, unlike PROTEP, POSSUM allows the use of more general partially-defined queries in which only a subset of the angles and/or distances between SSEs needs to be defined. This feature of POSSUM has proved especially useful in carrying out a wide-ranging survey of beta topologies and psi loops [74, 75], which showed that very few of the possible types of sheet were found to occur in practice, and also allowed the identification of several previously unidentified instances of psi loops.

The subgraph isomorphism algorithm used in POSSUM required all the secondary structure elements (SSEs) in the pattern to be present in a structure in the PDB if that structure was to be retrieved; partial matches between pattern and structure could not be found. The PROTEP program (PROtein Topographic Exploration Program) [53] uses the Bron and Kerbosch maximal common subgraph algorithm [71] to match the query nodes (in this case SSEs) to the structure nodes by looking at the relationship (graph edge values, within specified tolerances) between them. This permits the rapid location of any structural overlaps between the query structure and any of other proteins in the PDB, and output is interface to various molecular graphics programs. PROTEP was extensively tested against a variety of known motifs, including the trypsin, azurin and globin families, and shown to operate correctly and effectively [53]. Although computationally more intensive than POSSUM, because it allows partial matches to the query, PROTEP is still fast and a search of the entire PDB typically requires 10–30 min on an R4000 Silicon Graphics Indigo or DEC alpha AXP 3000/400 workstation, compared with 1–2 min for the simpler POSSUM procedure.

PROTEP has proved a very successful method for the detection of previously unrecognized similarities between families of protein structure with no obvious sequence homology [63–65]. In some cases these similarities may indicate a very distant common ancestry, as in the striking similarity we established between the families of Zn^{2+} aminopeptidases and carboxypeptidases [63]. In other cases the similarity is more likely to represent convergent evolution towards a stable structure: an example of this is the structural resemblance we have shown between the core beta sheets of the enzyme protocatechuate 3,4-dioxygenase and two other completely unrelated proteins, the intensely sweet protein thaumatin, and the thyroid-hormone transporting protein prealbumin [65]. We shall illustrate the use of PROTEP in such applications by discussing a further example in more detail in the next section.

4.2.1 Use of PROTEP: Structural Similarities Involving HIV Reverse Transcriptase

In the last year, using PROTEP, we have discovered very interesting and potentially medically important structural resemblances involving the enzyme reverse transcriptase (RT) from HIV virus [64]. HIV reverse transcriptase is a multifunctional enzyme of great medical importance because of its pivotal role in AIDS infection. RT catalyses the transcription of a single stranded retroviral RNA template into a single strand of DNA, and single stranded DNA into double-stranded DNA which is then subsequently incorporated into the host cell's genome [76]. This process is unique to retroviruses, and the RT of the HIV virus is therefore a prime target for potential anti-AIDS drugs such as AZT (3-azido-deoxythmidine) and ddI (dideoxyionisine), which both operate by incorrectly terminating the DNA strand as it is synthesized by RT [77]. However major medical problems have arisen because mutations in RT occur rapidly and confer resistance to these drugs [78]. The crystal structure of a complete HIV-1 RT at 3.5 Å resolution was determined by Kohlstaedt et al. [79] using X-ray diffraction. The molecule consists of two polypeptide chains of molecular weights 66 000 and 51 000, named "p66" and "p51" respectively. The p66 chain can be divided into five domains: the first three are called the "fingers", "palm" and "thumb" domains and grip the proposed position of the DNA/RNA duplex; the fourth domain is called the "connection" domain and leads into the RNaseH domain. The p51 chain is a truncated p66 chain: it lacks the RNaseH domain, but possesses the other four domains, although their relative positions are quite drastically altered.

The PROTEP program enabled us to identify two intriguing and previously unrecognized structural similarities involving HIV reverse transcriptase [64]. These are shown in Fig. 1. First, there is a resemblance between the RNaseH domain fold and the three-dimensional folds of the RNaseH domain of RT and the "ATPase folds" of enzymes such as hexokinase, the 70kD heat-shock cognate protein and actin. On comparing the RNaseH domain with the ATPase fold, it is clear that the five-stranded beta sheet in RNaseH and the five-stranded sheet in the ATPase fold are topologically identical (i.e., they share the same directionality and connectivity of beta strands and alpha helix-see Fig. 1) and overlap well in 3-D space. It can be argued that the RNaseH has a related function to ATPase fold molecules in that they are all involved in metal-dependent cleavage

residues	Strand 1 835-840		Strand 2 843-847		Strand 3 855-861		Helix 871-883		Strand 4 891-895		Strand 5 913-917	
	Å	°	Å	°	Å	°	Å	°	Å	°	Å	°
Strand 1	0	.0	160	4.7	-38	9.4	147	10.5	-15	4.5	-45	8.8
Strand 2	160	4.7	0	.0	162	4.7	-32	10.1	150	9.6	116	13.6
Strand 3	-38	9.4	162	4.7	0	.0	142	10.0	-47	14.0	-81	18.1
Helix	147	10.5	-32	10.1	142	10.0	0	.0	154	10.4	128	12.0
Strand 4	-15	4.5	150	9.6	-47	14.0	154	10.4	0	.0	-34	4.1
Strand 5	-45	8.8	116	13.6	-81	18.1	128	12.0	-34	4.1	0	.0

a

(i) RT connection domain

(ii) RNase H domain

(iii) actin Ia domain

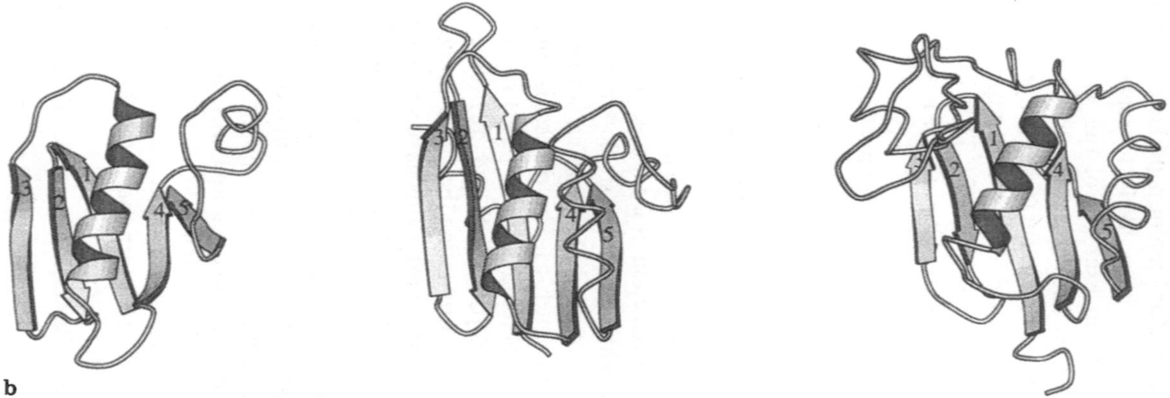

b

Fig. 1. a Matrix of distances and angles corresponding to the five strands and one helix in the RT connection domain (illustrated in Fig. 1b(i)). This is a subset of the actual matrix used in the search which consisted of 10 helices and 16 strands from the RT p51 domain. **b** Chain traces (produced with Molscript [80]) of (i) the RNaseH domain of RT; (ii) the connection domain of RT; and (iii) the Ia domain of actin. The helices and strands equivalenced in our study are represented as *coiled ribbons* and *sequentially numbered arrows* respectively; the non-equivalent parts of the structure are shown as a *smoothed C-alpha trace*

of a phosphodiester bond. It is not impossible that this similarity may indicate the existence of an ancient evolutionary connection between these classes of molecule. Nevertheless, a note of caution must be sounded: it is quite possible that the similarities between RNaseH and the ATPase fold are due to the processes of convergent evolution towards similar structures that catalyse similar reactions.

However the situation becomes more intriguing when one considers the second structural similarity that was detected. It became apparent that the motif that is common to RNaseH and the ATPase fold proteins is also present in the RT connection domain (see Fig. 1). Sequence comparisons reveal no significant similarity between sequences of the RNaseH and connection domains of RT, and thus there can be no direct evidence of any ancestral gene duplication event in the evolution of the RT protein. However it is possible that this structural similarity may indicate the presence of a previously unsuspected ribonuclease enzymatic site in the RT structure. In particular, we were able to propose [64] that the p51 connection domain may be the location of the RNase site which is responsible for removing the tRNA primer from the growing strand of DNA [81], and may therefore represent a further target for the design of therapeutic agents against HIV infection.

Therefore, in view of our present lack of knowledge concerning this important therapeutic target, the structural comparison was able to reveal a number of new and potentially important aspects of the structure of RT. These related both to possible evolutionary origins of some of the functions of RT, and to potential candidates for the locations of additional functional sites involved either in catalysis or in stabilization of nucleic acid.

5 Investigation of Local Protein Structure

Although the comparisons of protein folds can yield valuable insights into protein function and evolution, it is also very desirable to be able to detect structural similarities at the residue or atomic level. This is because the detection of similar patterns of functional groups in different proteins may allow analogies to be drawn between disparate proteins' modes of action. The classic example of this are the similar clusters of three catalytic residues in the otherwise unrelated subtilisin and chymotrypsin families of enzymes [82].

As with the detection of similarities in protein folds, research in this area has been intensive in recent years. One approach has been to use commercial relational database management systems (RDBMSs) in order to store and organize information about protein structure. Typically, information concerning atomic contacts, conformational angles, protein names etc. are held in tables and the database is then interrogated using a query language. Examples of such database systems include BIPED [83] and BRUGHEL [84]. Valuable analyses conducted using such systems include analyses of water structure in proteins and analyses of sidechain contacts [85, 86]. Other researchers consider that commercial RDBMS databases are not entirely appropriate to the specific problems of conducting searchers on protein structures. Thus Gardner and Thornton [87] observe that one fundamental problem is that, while proteins consist of serially linked amino acids, there is no easy way to access the implicit order of records in a RDBMS; they therefore developed their own order-based database program called IDITIS, which interacts with a program shell which will be able to incorporate such functions as fuzzy sequence, torsion angle and secondary structural searches. Using these techniques, several investigations have been carried out on specific types of interactions between sidechains which have involved pairs of aromatic residues [85, 88], ion-pairs and aromatic and amino groups [89, 90]. These studies have focused on analysing the details of the type of contact formed in order to determine preferred orientations of the pairs of sidechains. Furthermore, from the identification of clear packing preferences, valuable insights have been gained into the forces of interaction in these complex systems. Using their suite of programs SIRIUS, embedded in the IDITIS relational database of protein structure, Singh and Thornton [91] have compared the geometry of all interacting pairs of residues in the database. This has produced a three-dimensional distribution defining the spatial disposition and relative orientation of the interacting pairs of sidechains. This pairs distribution provides a powerful means of assessing preferred packing arrangements in protein structures. In addition, larger cluster of residues can be identified by combining pairs of queries [82].

Non-database approaches have also been used to identify similar patterns of sidechains in proteins. Bachar et al. [48] and Fischer et al. [49] discuss the matching of protein structures using geometric-hashing techniques that were first developed in research on computer vision. Another approach has been adopted by Vedani et al. [92], who introduced the concept of using anchor points from a set of known ligands to derive a pseudo-receptor model of residues at specified distances; given an appropriate database-searching algorithm, it would then be possible to identify proteins containing equivalent residue patterns for consideration as potential receptor sites.

Another program that considers the orientation of sidechains, albeit in a very different manner, is the CAVEAT program that has been developed by Bartlett and co-workers [93]. CAVEAT is used to search a database of 3-D small molecules, normally the Cambridge Structural Database [30], to identify those that have bonds that have the same geometric relationship to each other as C-alpha to C-beta bonds in a protein target. Vectors are drawn along the bonds of interest and then the geometric relationships between pairs of bonds are defined in terms of angles and distances between the corresponding vectors. Gregory et al. [94] have described a way of generating query templates using structural information from known metal-binding sites in proteins; they then use the two-stage pattern matching approach of Brint et al. [51] to search for these templates in the PDB, with the aim of identifying further potential metal-binding sites.

5.1 Comparison of Proteins at the Sidechain Level
Using Algorithms Derived from Graph Theory

5.1.1 Pseudo-Atom Representation of Sidechains

In Sect. 4.2 we have described the use of subgraph-matching algorithms for the rapid comparison of protein structures at the level of alpha helices and beta strands [52, 53]. The ASSAM program extends the methods that we developed previously to the representation and searching of arrangements of protein sidechains in 3-D space. In our earlier studies, a 3-D structure was represented by a graph in which the nodes were vectors that described the alpha helix and beta strand secondary structure elements, and in which the edges described the angles and distances between pairs of these vectors. In the ASSAM work, the nodes of the graph are the atoms in the sidechains, and the edges are the inter-atomic distances. In principle, all of the atoms in each sidechain could be used in the representation, but in practice there are a number of problems involved in defining equivalent constellations of residues in different proteins at this detailed level. This is because a significant degree of structural variation can occur even between different coordinate sets representing the same or closely related proteins. Thus equivalent sidechains may have different torsion angles or positions in different coordinate depositions. This can have a variety of causes including movement of sidechains as a result of binding substrates, disorder in sidechain positions in one or more structures, and ambiguities in positioning atoms at the ends of sidechains (e.g., the positions of the nitrogen and oxygen atoms of the amide group at the end of asparagine, or the final torsion angle of valine or leucine), especially in medium-resolution protein crystallographic studies.

To some degree these structural variations can also be allowed for by appropriate choice of the inter-atomic distance tolerances that are specified when a query pattern is to be searched against the PDB. However, it became clear that ambiguities of the kind described above could be accommodated by using a simplified representation of the sidechain positions, in which each sidechain is characterised by a small number of pseudo-atoms. This approach not only enhances the flexibility of the query procedure, but also improves the speed of searching because of the smaller number of nodes that need to be considered in the graph-matching operations when compared with a representation involving all of the atoms in a sidechain. The latter is of particular importance in the context of a database-searching program, since a subgraph isomorphism algorithm must, of necessity, have a running time that is proportional to the factorials of the numbers of nodes in the graphs that are being matched. We therefore define the relative orientations of sidechains in space by means of distances (the edges of the graphs) between pseudo-atoms representing the sidechain (the nodes of the graph). In practice, each sidechain is represented by just two pseudo-atoms, one near to the start and the other near to the end of the functional part of the sidechain; here we shall refer to these as the "S" and "E" pseudo-atoms, respectively. Some of our searches have also made use of a third, midpoint position (denoted by the letter "M"), which is calculated as the mean of the S and E positions. This pseudo-atom representation of a protein is a graph in which the nodes are the vectors joining the pseudo-atoms and the edges are the set of distances (SS, SE, MM, ES and EE) between pairs of these vectors. Thus the pair of pseudo-atoms that describe a particular sidechain may usefully be considered as a vector, so that the nodes of the graphs considered here are the vectors describing the presence of a sidechain (as against the presence of an SSE as in our previous work). In fact, the nodes considered here are examples of labelled graphs, since there are labels associated with the nodes and edges, namely the residue types and the set of value for the inter-vector distances, respectively.

The pseudo-atom positions used here for each of the sidechain types are illustrated in Fig. 2. Residues with similar chemical properties are represented analogously. Thus, for example, one of the pseudo-atoms for glutamate is placed on the CG position, and the other the mean of the OE1 and OE2 positions; for aspartate the pseudo-atoms are equivalently positioned on CB and on the mean of OD1 and OD2 respectively. This means that glutamate and aspartate residues can be compared directly, if desired, as the S and E points are in equivalent positions in the two carboxyl groups, thus enabling the specification of a generic acid group that corresponds to both glutamate and aspartate. In addition to the 20 standard amino acids, it is possible to specify the following generic amino acid types in a search: acidic, amide Asx, Glx, basic, aromatic, small and large hydrophobic. At present the emphasis on the positioning of the pseudo-atoms is on the functional part of the residue; for example, on the carboxyl group in Asp and Glu and on the guanidinium group in Arg. However the particular number and distribution of pseudo-atoms used to represent each sidechain can easily be changed in future, if desired, or it would be possible to use the positions of actual atoms themselves.

Thus, in our current implementation of the method, we represent the position of a sidechain by the two pseudo-atoms, and the relative orientations of pairs of sidechains by the distances between them. Specifically, for each pair of sidechains in a query pattern (or in a database structure), five distances are calculated, these being the SS, SE, EE, ES and MM distances as illustrated schematically in Fig. 3, which shows a pattern of three residues and the associated inter-atomic distances. Although these five distances (or a user-defined subset of them) provide a very simple way of defining the orientation of a pair of sidechains, our investigations (as detailed below) indicate that the distances are extremely effective in detecting similarities (both known and previously unknown) between sidechain arrangements.

5.1.2 The ASSAM Search Program

In order to conduct a search, the graphs that represent each of the proteins in the PDB must be examined using a subgraph isomorphism algorithm to see if they

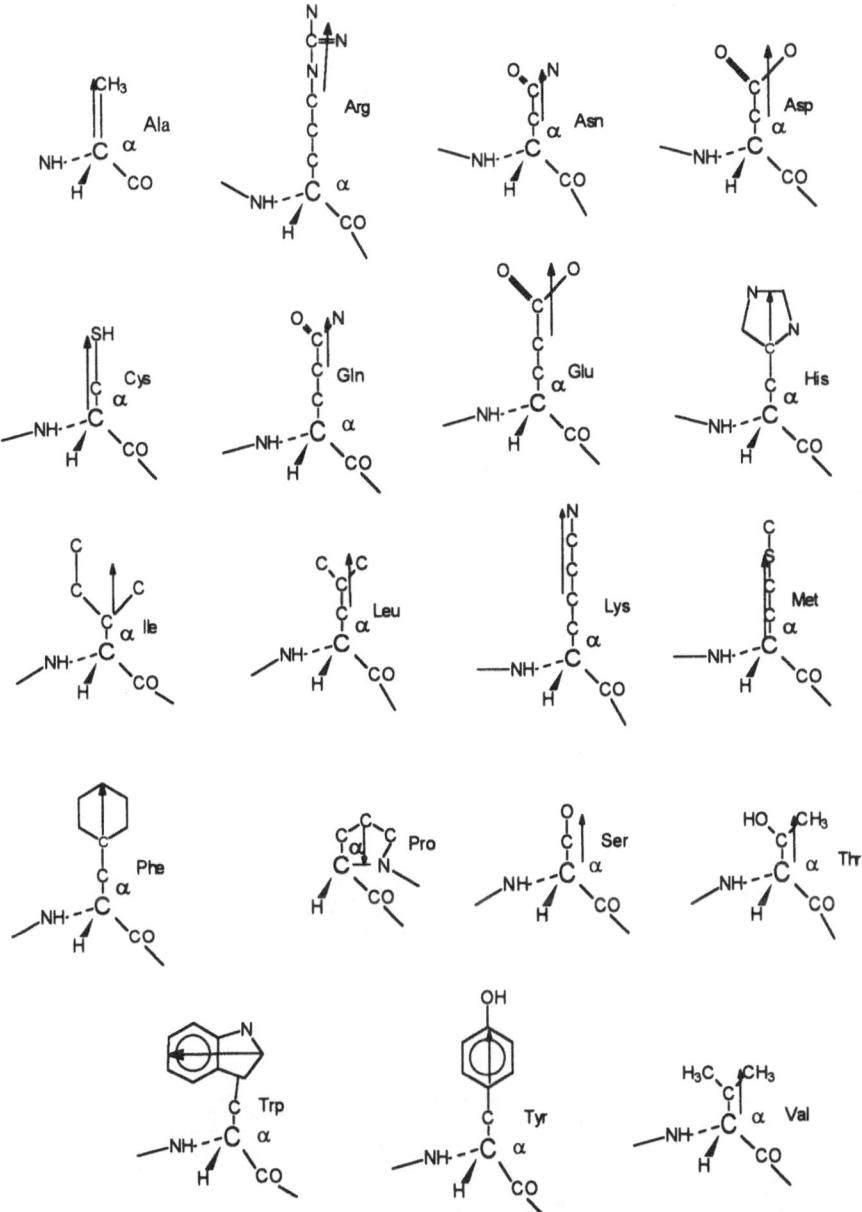

Fig. 2. The pseudo-atom positions used by the ASSAM program for the representation of each of the 20 different sidechain types that occur in normal proteins. The "start" (S) and "end" (E) pseudo-atoms are situated at the beginning and end, respectively, of the arrows on the diagram and delineate the functional part of the sidechain

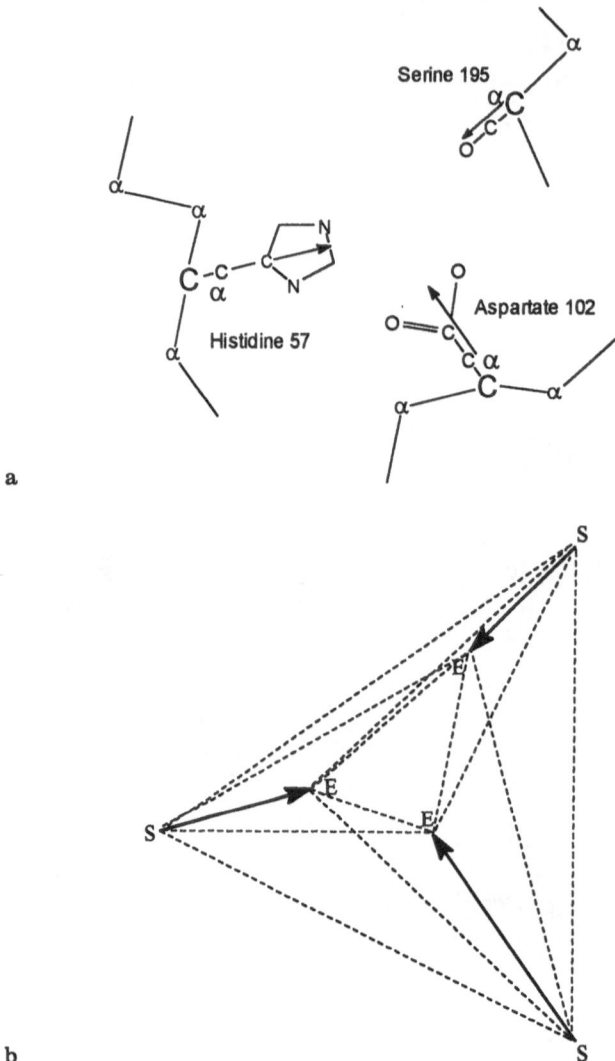

a

b

	His57				Asp102				Ser195			
His57	0	0	0	0	5.7	3.6	6.3	4.3	6.4	5.0	4.4	3.1
Asp102	5.7	6.3	3.6	4.3	0	0	0	0	9.3	9.0	7.7	7.2
Ser195	6.4	4.4	5.0	3.1	9.3	7.7	9.0	7.2	0	0	0	0

c

Fig. 3a–c. Schematic representation of the three residues involved in the formation of the catalytic triad of chymotrypsin [82] with the pseudo-atom vectors (*arrows*) overlaid: **a** the positions of real non-hydrogen atoms in the sidechains, residues; **b** the S and E pseudo-atom positions with the associated distances between them (*dotted lines*) (for clarity the MM distances are not shown); **c** the search matrix for this grouping of sidechains, the distances, all of which are in Å, being for the SS, SE, ES and EE distances respectively

contain a subgraph equivalent to that defined in the query matrix. The graphs that represent the structures in the PDB are not stored directly as such; instead the necessary parts are generated as required for a particular search pattern. Specifically, each protein has an associated text file that contains the coordinates of the pseudo-atoms for each of its constituent amino acids. When a query is matched against a protein, the pseudo-atoms coordinate information is used to generate the inter-vector distances only for those pseudo-atom types and distance types that have been specified in the query. The calculation of the distances "on the fly" minimises the storage overheads, and greatly reduces the input-output overheads that would be involved in reading large pre-calculated matrices.

A search pattern is defined by specifying a set of residues from the coordinate set of a protein in the PDB, and the search matrix containing the query vectors and inter-vector distances is then generated automatically. The user is prompted for the tolerances that are to be applied to each of the query distances, and for any generic amino acid types that are to be employed in the search. Each PDB structure in turn is read in from disk, and then the matrix for the appropriate amino acids calculated as described above. The query matrix is sought in the protein matrix using the Ullmann algorithm [70], and hits output for subsequent inspection. The program has been implemented in FORTRAN 77 on Silicon Graphics and Evans and Sutherland workstations running under the UNIX operating system.

To test the program, patterns were constructed using commonly occurring or well-characterized clusters of residues, so that it would be possible to compare the matches identified by the program with those one would expect to find. A pattern constructed from the Asp-His-Ser catalytic triad residues from alpha-chymotrypsin [82] was used to test the program in this way: in addition to correctly retrieving the known instances of catalytic triad the program detected a surprising and intriguing instance of a second Asp-His-Ser triad in certain chymotrpysinogen and trypsinogen structures [66]. Test searches were also carried out [66] on two other PDB-derived patterns: a pair of arginines from the staphylococcal nuclease catalytic site [95] which are important in hydrogen-bonding the phosphate of the diester substrate, and are thought to activate it to nucleophilic attack [95]. This pattern showed an interesting similarity to a pair of arginine sidechains in thymidylate synthase [96] that are hydrogen bonded to the phosphate group of the UMP [66]. All these searches were carried out on a 791-structure subset of the April 1993 release of the PDB and required less than 3 min of real and CPU time on a Silicon Graphics R4000 Indigo workstation. Another search conducted [66], using just MM distances, was on a pattern consisting of zinc-coordinating sidechains from thermolysin [97] which showed an expected similarity to the zinc-coordinating sidechains of carboxypeptidase [98]. This last example is of particular interest because, although unrelated in sequence and in overall structure, it is known that there are similarities in the active sites of thermolysin and carboxypeptidase A [97] and that both bind phosphoramidate inhibitors [99]. The latter fact suggests that the ability to detect similarities between binding sites may be a useful indicator of analogous binding properties, and may therefore have potential applications in the area of rational molecular design, since structural similarity between a site in protein "A" and a known ligand-binding site in another unrelated protein, "B" might suggest that a derivative of B's ligand may be capable of binding to A. A possible example of this, detected using ASSAM, is discussed in the next section.

5.1.3 Structural Similarity Between Binding Sites in Influenza Sialidase and Isocitrate Dehydrogenase

Recently, von Itzstein et al. [100] described the design of potential anti-influenza drugs which operate by binding to a specific site on the influenza sialidase molecule, thereby inhibiting it. This was achieved using a combination of computer-assisted manual examination of the active site structure with bound sialic acid and sialic acid analogues [101, 102], together with the use of the GRID program [103] to explore probable interaction sites between probes with various functional group characteristics and the enzyme surface. A key feature of their analysis of the sialidase structure was the recognition of a negatively charged patch on the sialidase surface provided by two glutamate residues and not used in the binding of sialic acid. This patch was exploited in the rational design process by the synthesis of a sialic acid analogue bearing an additional positive charge from either an amino- or a guanidinyl-group appropriately placed to interact with this region of the sialidase surface. This work constituted a long-awaited example of rational computer-assisted design of a new drug based on the crystal structure of a target protein [104], and highlighted the great importance of knowing the three-dimensional structures of medically important macromolecules.

von Itzstein et al. identified a group of four sidechains as being involved in drug-binding in the influenza sialidase (two arginines, 371 and 118, and two carboxylic acid groups, glutamates 119 and 227). Coordinates for these sidechains were taken from the sialidase structure [105] and were used to generate a search pattern corresponding to this sidechain cluster. Using the ASSAM program we compared this three-dimensional arrangement of sidechains with all the other protein structures in the Protein Data Bank (PDB (October 1992 release)). The search revealed [106] that there is a very similar cluster of four sidechains in the binding pocket for the isocitrate/Mg^{2+} complex in the active site of E. coli isocitrate dehydrogenase (ICDH) [107], an enzyme which catalyses the $NADP^+$-linked oxidative decarboxylation of isocitrate to 2-oxoglutarate.

The two sets of sidechains from sialidase and from ICDH are shown superposed in Fig. 4 where it can be seen that in ICDH the positions of arginines 119 and 153 correspond closely to the two arginines 371 and 118 respectively in the sialidase, with excellent overlap of the guanidinium groups. In ICDH these two arginines interact with two carboxyl groups of the isocitrate, one of which, the C3 carboxyl, occupies an equivalent position to the carboxyl group of the inhibitor 4-guanidino-Neu5Ac2en in sialidase. The two carboxyl groups in the

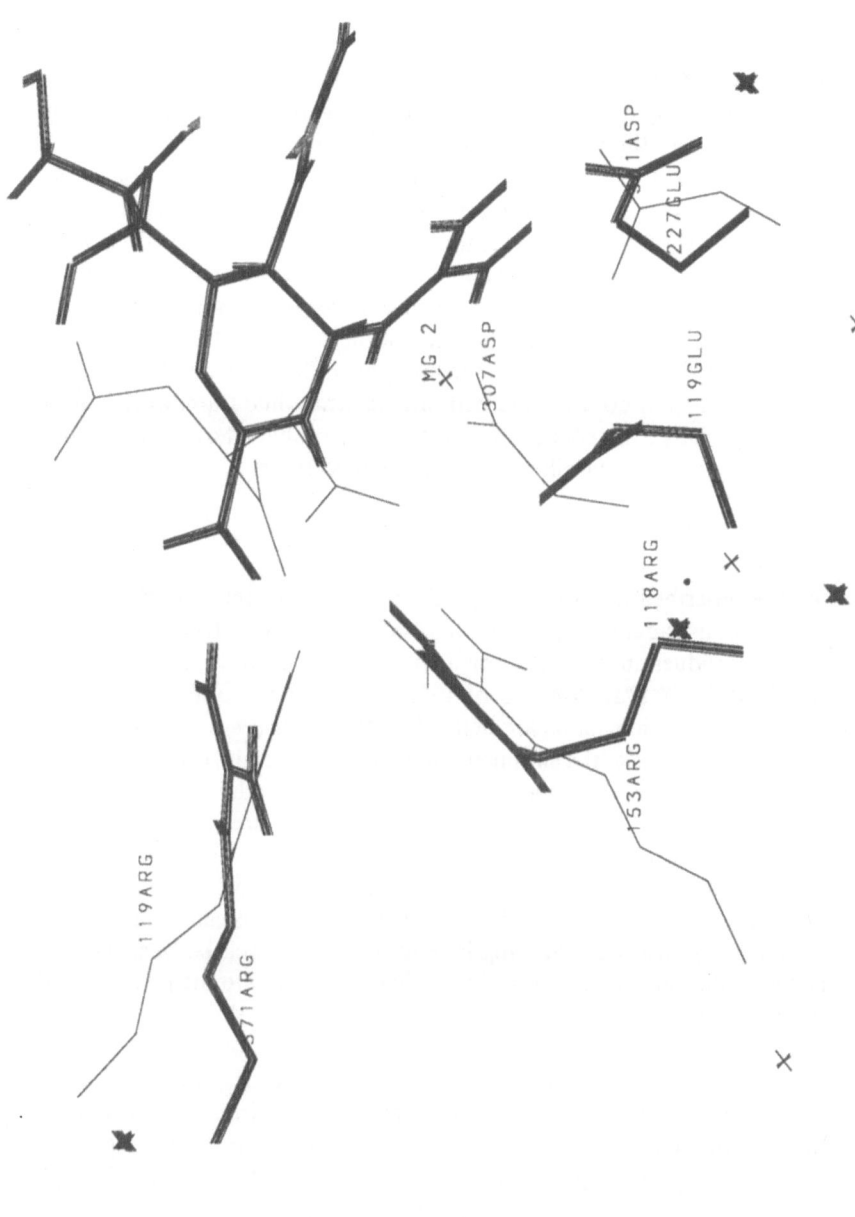

Fig. 4. Superposition of the sialidase and ICDH binding sites. Arginines 371 and 118 and glutamates 119 and 227 from sialidase are shown in *bold lines*, as in the inhibitor 4-guanidino-Neu5Ac2en (*upper right*). Arginines 119 and 153 and aspartates 307 and 311 of ICDH, together with iscocitrate and magnesium ion, are shown *superposed* on the sialidase in *light lines*

ICDH site are provided by two aspartates (aspartates 307 and 311) which are slightly displaced with respect to their counterparts in sialidase (glutamates 119 and 227 respectively). In sialidase these carboxyls coordinate the positive charge of the 4-guanidinyl group of the inhibitor, and in ICDH the equivalent carboxyls analogously coordinate the positive charge of the Mg^{2+} ion in the complex.

Clearly this result does not mean that the sialidase would necessarily be expected to bind isocitrate. The binding site for sialic acid in sialidase comprises many more residues than the two arginines and two aspartates that were highlighted by von Itzstein et al. [100] and which we used as our search pattern. In fact these four residues constitute the only major area of resemblance between the binding sites in the two proteins; the folds of the polypeptide chain and the positions of other residues in the active site are quite different. Nevertheless, this finding does suggest that compounds with an isocitrate-like framework and a similar pattern of charge distribution might form a suitable alternative starting point for the design of new families of anti-infleunza drugs.

5.1.4 Future Directions

Above we have described the use of graph theoretical algorithms to identify all occurrences of a user-defined pattern of residues in a database of protein structures. The residues in a protein or a query pattern are represented in a highly simplified form that consists of two pseudo-atoms and the relative orientations of pairs of sidechains are defined by the distances between pairs of these pseudo-atoms. Despite the simplicity of the representation, our tests with a variety of patterns demonstrate the usefulness of the methodology.

The graphs that form the basis for the search program are very simple, and use only inter-vector distance information to specify the geometric relationships between pairs of residues. It would be of interest to determine whether the use of inter-vector angles could improve the retrieval performance still further. In addition, where more precise searching is required it may well be beneficial to adopt a less minimal sidechain description, either using real atomic positions or increasing the number of pseudo-atoms. It is also clear that the procedure will benefit from a number of extensions in the structural description, notably the addition of mainchain atoms, and the incorporation of a description of non-protein structural elements including nucleic acid, carbohydrate, substrate and ion positions. Although such extensions will be valuable, the studies reported here show that the present procedure, with its simplified representation of sidechains, has very considerable merits, notably in terms of flexibility of query definition and of speed of use, as a method for the comparison of sidechain clusters in three dimensions.

5.2 Other Applications

The publication of Mitchell et al. (1990) [52] led to interest in other groups in the use of graph theoretical methods for protein structure comparison. Notable amongst these was the group of the late I. Hanneef: Subbarao and Haneef [54] described the use of algorithms derived from graph theory in order to define topologically equivalent residues in marcomolecules, the equivalences being assigned according to the environments (defined in terms of inter-residue distances) of such residues. For these purposes they used the Ullmann subgraph isomorphism algorithm [70], and also a modified version of the algorithm described by Bersohn et al. [108] in order to search for maximal common subgraphs in those cases where it was not possible to predefine a search pattern. Using these methods they showed that it was possible to align the structures of different protein molecules at the alpha-carbon level of description, or of different RNA molecules at the backbone level of description and showed that they could obtain similar results for specific comparisons to those reported in the literature by earlier workers [54]. A comparison of the small protein dihydrofolate reductase with a reduced database of 107 3-D structures required 26 h of cpu time on a microVAX 3600. In addition, Subbarao and Haneef described the use of their algorithm for scanning the (1-D) sequence database for sequences homologous to a search pattern [54]. In a further application of graph theoretical methods to sequence searching, Heringa and Argos [109] have described an algorithm for delineating similarities between fragments of protein sequences which uses graph theoretical approaches to delineate start and end points of the sequence fragments. More recently, Haneef and coworkers [110] built on their earlier work [54] and extended their graph theoretical methodology so as to address the problem of the docking together of the complementary surfaces of molecules. This is a similar problem to similarity searching except that the requirement is to detect structural complementarity rather than direct similarity. The use of the method was demonstrated in examples involving antibody-antigen recognition and its potential use in drug design was discussed [110].

Koch et al. [111] have discussed the use of graph theoretical techniques in an attempt to find rules to relate beta sheet topology to amino acid sequence and for the comparison of beta sheet structures. They defined a graph representation for every protein in the PDB that contains beta sheets, notations and graphic representations for sheets which described the sequential and topological neighbourhoods of the strands, and constructed tools for substructure searches of this database.

Graph theoretical algorithms also play a part in the methodology described by Rufino and Blundell [112] who use an SSE-based description of protein structures in order to permit the detection of structural relationships. In their methodology, the largest set of equivalent secondary structures is determined using the maximal common subgraph algorithm of Bron and Kerbosch [71], and an elaborate scoring mechanism coupled with a dynamic programming algorithm is then used in order to evaluate the similarity and to construct phylogenic trees of the inter-relatedness of protein families [112].

6 Conclusions

An important factor in the development of molecular biology has been the availability of computational tools for the detection of similarities in the one-dimensional sequences of proteins and nucleic acids [12]. It is similarly of the greatest importance to be able to detect structural analogies within the rapidly growing database of three-dimensional protein and nucleic acid structures in order to enhance understanding of structure/function relationships in biological macromolecules. The reverse transcriptase fold similarities discussed in Sect. 4.2.1, and the influenza sialidase binding site sample discussed in Sect. 5.1.3, suggest that such comparative studies may give valuable insights into structures of medical interest, and represent another valuable approach to the rational design of novel inhibitors.

In more general terms, it is clear that structural comparisons of the kind reported here may be more widely applicable. At present the database of known three-dimensional protein structures is very small, consisting of only a few hundred distinct structures. As this database expands, it is increasingly likely that more similarities of the kind we have described above will be observed between otherwise disparate proteins. At the scientific level, such similarities will enhance our understanding of structure/function relationships in proteins, while technologically they may prove valuable both in protein engineering and in the search for new lead compounds in drug design. Thus it seems likely that structural comparisons will be a useful weapon to be used in conjunction with other modelling procedures.

The last twenty years have seen an exponential growth in the information available concerning the sequences of biological macromolecules. All the indications are that this expansion will continue: enterprises such as the Human Genome Project (Sect. 2.1.1) will ultimately make complete our knowledge of human genetic and make available the sequences of all human proteins. In fact, the ultimate goal of many of the crystallographic and comparative studies on protein structure is to learn how to predict a protein's 3-D structure from its sequence alone (see Sect 2.1.1). However, the major problem at present with all predictive methods is that the knowledge base is too small to allow the derivation of rules for protein folding [113], and this is reflected in the failure so far of both human "experts" and computers to predict previously unknown protein structures. It is hoped that the explanation of the structural similarities between proteins of disparate sequences will lead to a greater understanding of this important problem.

However, just as our knowledge of protein sequences has increased in recent years, an analogous growth, albeit on a smaller scale, has taken place in the extent of our knowledge of the three-dimensional structures of biological marcromolecules, and these studies have been massively accelerated by the availability of formerly rare proteins overexpressed using the techniques of genetic engineering. Computational methods which make possible the comparison of sequences and structures have been, and will continue to be, crucial to the success of these enterprises. It is fortunate that, in addition to the developments of new theoretical methodologies, great improvements in the power of computers have taken place over the timescale of these developments in molecular biology.

Acknowledgments. We gratefully acknowledge the UK Science and Engineering Research Council, the Lister Institute, the Wellcome Trust and Tripos Associates Inc. for hardware and support. PJA is a Royal Society University Research Fellow. The Krebs Institute is a designated centre of the SERC Biomolecular Science. We are very grateful to Andrew Brint, Eleanor Mitchell, Helen Grindley, Elizabeth Ujah, Julie Park, Kiran Kumar and Amanda Mackenzie, all of whom have made contributions to this work.

7 References

1. Ash JE, Warr WA, Willett P (eds) (1991) Chemical Structure Systems. Ellis Horwood, Chichester
2. Watson JD, Crick FHC (1953) Nature 171: 371
3. Hunkerpillar MW, Strickler JE, Wilson KJ (1984) Science 226: 304
4. Nirenberg M (1973). In: Nobel Lectures: Physiology or Medicine. American Elsevier, pp, 272–395
5. Sanger F, Nicklen S, Coulson AR (1977) Proc Natl Acad Sci USA 7: 5463
6. Maxam A, Gilbert W (1977) Proc Natl Acad Sci USA 74: 560
7. Sweeley CC, Nunez H (1985) Ann. Rev, Biochem. 54: 765
8. Bleasby AJ, Wooton JC, Akrigg, D, Dix NIM, Findlay JBC, North ACT, Parry-Smith D, Islam S, Gardner SP, Thornton JM, Sternberg MJE, Blundell TL, Hayes FRF & Tickle IJ (1988) ISIS. Integrated Sequence/Integrated Structure Resource, British Biotechnology Group
9. Lesk, AM (1988) in: Lesk AM (ed) The EMBL Data Library" in Computational Molecular Biology, Oxford University Press, Oxford, pp 55–65
10. Watson JD (1990) Science 44: 44
11. Cantor CR (1990) Science 44: 49
12. Lesk AM (ed) (1988) Computational Molecular Biology Oxford University Press, Oxford
13. Dayhoff MM (1978) in: Atlas of Protein Sequence and Structure, National Biomedical Research Foundation, vol 5, suppl 3
14. Needleman SB, Wunsch CD (1970) J Mol Biol 48 443
15. Staden R (1982) Nucleic Acids Res 10: 2951
16. Maizel JV and Lenk RP (1981) Proc Natl Acad Sci USA 78: 7665
17. Unger R, Harel D, Sussman JL, (1986) Comp Appl Biosci 2: 283
18. Smith TF and Waterman MS (1981) Adv Appl Math 2: 482
19. Lipman DJ, Pearson WR (1985) Science 227: 1435
20. Wilbur WJ, Lipman DJ (1983) Proc Natl Acad Sci USA 80: 726
21. Deisenhofer J, Epp O, Miki K, Huber R, Michel H (1985) Nature 318: 618
22. Jurnak FA, McPherson A (eds) (1984) "Biological macromolecule and assemblies. Volume 1: Virus Structure", Wiley.
23. Blundell TL, Johnson LN (1976) "Protein Crystallography", Academic Press, London
24. Glusker JP, Trueblook KN (1972) "Crystal structure analysis: A primer", Oxford University Press, Oxford
25. Wüthrich K (1986) "NMR of Proteins and Nucleic Acids", Wiley, New York
26. Henderson R, Baldwin JM, Ceska TA, Zemlin F, Beckmann E, Downing KH (1990) J. Molec. Biol. 213: 899
27. Kühlbrandt W, Wang DA, Fujiyoshi Y (1994) Nature 367: 614

28. Bernstein FC, Koetzle TF, Williams GJB, Meyer EF Jnr., Brice MD, Rodgers JR, Kennard O, Shimanouchi M, Tasumi M (1977) J Molec Biol 112: 535
29. Abola EE, Bernstein FC, Bryant SH, Koetzle TF, Weng J (1987) in: Allen FH, Bergeroff G, Sievers R (eds) Crystallographic Databases-Information Content, Software Systems, Scientific Applications pp 107–132, Data Commission of the International Union of Crystallography, Bonn/Cambridge/Chester
30. Allan FH, Kennard O, Taylor (1983) Acc. Chem. Res. 16: 146
31. Richards FM, and 173 others (1988) "Letter to the editor" J Molec. Graph. 6: 178
32. Hendrickson W (1985) in: Wyckoff HW, Hirs CHW, Timasheff SN (eds) pp 252–270 (Methods in Enzymology, vol 115)
33. Colman PM, Deisenhofer J, Huber R, Palm W (1976) J Mol Biol 100: 257
34. Richards FM (1991) Scientific American 264: 54
35. Chothia C (1988) Nature 333: 598
36. Rao ST, Rossmann MG (1973) J Mol Biol 76: 241
37. Rossmann MG, Argos P (1975) J Mol Biol Chem 250: 7525
38. Rossmann MG, Argos P (1976) J Mol Biol 105: 75
39. Rossmann MG, Argos P (1977) J Mol Biol 109: 99
40. Remington SJ, Matthews BW (1978) Proc Nat Acad Sci USA 75: 2180
41. Remington SJ, Mathews BW (1980) J Mol Biol 140: 77
42. Lesk AM (1979) Comm ACM 22: 219
43. Lesk AM (1993) J Chem Soc Farad Trans 89: 2603
44. Taylor WR, Orengo CA (1989) J Mol Biol 208: 1
45. Sali A, Blundell TL (1990) J Mol Biol 212: 403
46. Vriend G, Sander C (1991) Proteins: Struct Funct and Genet 11: 52
47. Alexandrov NN, Takahashi K, Go N (1992) J Mol Biol 225: 5
48. Bachar O, Fischer D, Nussinov R, Wolfson HJ (1993) Prot Eng 6: 279
49. Fischer D, Bachar O, Nussinov R, Wolfson H (1992) J Biomol Struct Dynam 9: 769
50. May ACW, Johnson MS (1994) Prot Eng 7: 475
51. Brint AT, Davies HM, Mitchell EM, Willett P (1989) J Mol Graph 7: 48
52. Mitchell EM, Artymiuk PJ, Rice DW, Willett P (1990) J Mol Biol 212: 151
53. Grindley HM, Artymiuk PJ, Rice DW, Willett P (1993) J Mol Biol 229: 707
54. Subbarao N, Haneef I (1991) Prot. Eng. 4: 877
55. Goodsell DS, Lauble H, Stout CD, Olson AJ (1993) Protiens: Struct Funct Genet 17: 1
56. Smellie AS, Crippen GM, Richards WG (1991) J Chem Inf Comput Sci 31: 386
57. Richards FM, Kundrot CE (1988) Proteins: Struct Funct Genet 3: 71
58. Willett P (1991) Three-Dimensional Chemical Structure Handling, Research Studies Press, Wiley, New York
59. Willett P (ed) Modern Approaches to Chemical Reaction Searching, Gower, Aldershot
60. Hagadone TR (1992) J Chem Inf Comp Sci 32: 515
61. Artymiuk PJ, Rice DW, Mitchell EM, Willett P (1990) Prot Eng 4: 39
62. Artymiuk PJ, Rice DW (1991) in: Ash JE, Warr WA, Willett P (eds) Chemical Structure Systems, Ellis Horwood, Chichester, pp 299–328
63. Artymiuk PJ, Grindley HM, Park JE, Rice DW, Willett P (1992) FEBS Lett 303: 48
64. Artymiuk PJ, Grindley HM, Kumar K, Rice DW, Willett P (1993) FEBS Lett 324: 15
65. Artymiuk PJ, Grindley HM, MacKenzie AB, Rice DW, Ujah EC, Willett P (1994) in: Carbo R (ed) Molecular Similarity and Reactivity: from quantum chemical to phenomenological approach, Klure Academic Press (in the press)
66. Artymiuk PJ, Poirrette AR, Grindley HM, Rice DW and Willett P (1994) J Mol Biol 243: 327–344
67. Kabsch W, Sander C (1983) Biopolymers 22: 2577
68. Brint AT, Willett P (1987) J Mol Graph 5: 49
69. Brint AT, Willett P (1987) J Chem Inf Comput Sci 27: 152
70. Ullmann JR (1976) J.ACM 16: 31
71. Bron C, Kerbosch J (1973) Comm ACM 16: 575
72. Barrow HG, Burstall RM (1976) Inf. Proc. Lett. 4: 83
73. Richardson JS (1977) Nature 268: 495
74. Artymiuk PJ, Grindley HM, Poirrette AR, Rice DW, Ujah EC, Willett P (1994) J Chem Inf Comput Sci 34: 54
75. Ujah EC (1992) Study of beta-sheet motifs at different levels of structural abstraction using

graph-theoretic and dynamic programming techniques. PhD Thesis, University of Sheffield, Sheffiled

76. Goff SP (1990) J AIDS 3: 817
77. Mitsuya H, Yarchoan R, Broder S (1990) Science 249: 1533
78. Wlodawer A (1992) Science 256: 1766
79. Kohlsatedt LA, Wang J, Friedman JM, Rice PA and Steitz TA (1992) Science 256: 1783
80. Kraulis PJ (1991) J Appl Cryst 24: 946
81. Peliska JA, Bencovic SJ (1992) Science 258: 1112
82. Blow DM (1990) Nature, 343: 694
83. Islam SA, Sternberg MJE (1989) Prot Eng 2: 431
84. Delhaise P, Bardiaux M, Wodak S (1985) J Mol. Graph. 3: 116
85. Singh J, Thornton JM (1985) FEBS Lett. 191: 1
86. Sing J, Thornton JM, Snarey M, Campbell SF (1987) FEBS Lett 224: 161
87. Gardner SJ, Thornton JM (1990) in: Abstracts of International Conference on Computing in Molecular Biology, Chester UK
88. Burley SK, Petsko GA (1985) Science 229: 23
89. Burley SK, Petsko GA (1988) Adv Prot Chem 39: 1215
90. Singh J, Thornton JM, Snarey M, Campbell SF (1987) FEBS Lett. 224: 161
91. Singh J, Thornton JM (1990) J Mol Biol 211: 595
92. Vedani A, Zbinden P, Snyder JP (1993) J Receptor Res. 13: 163
93. Bartlett PA, Shea GT, Telfer SJ, Waterman S (1990) in: Roberts SM (ed) Molecular Recognition: Chemical and Biochemical Problems, Royal Society of Chemistry, Cambridge, pp. 182–196
94. Gregory DS, Martin ACR, Cheetham JC, Rees AR (1993) Prot Eng 6: 29
95. Cotton FA, Hazen EE, Legg MJ (1979) Proc Natl Acad Sci USA 76: 2551
96. Montfort WR, Perry KM, Fauman EB, Finer-Moore JS, Maley GF, Hardy L, Maley F, Stroud RM (1990) Biochemistry 29: 6964
97. Kester WR, Matthews BW (1977) J Biol Chem 252: 7704
98. Rees DC, Lewis M, Lipscomb WN (1983) J Mol Biol 168: 367
99. Kam C-M, Nishino N, Powers JC (1979) Biochemistry 18: 3032
100. von Itzstein M, Wu W-Y, Kok GB, Pegg MS, Dyason JC, Jin B, Phan TV, Smythe ML, White HF, Oliver SW, Colman PM, Varghese JN, Ryan DM, Woods JM, Bethell RC, Hotham VJ, Cameron JM, Penn CR (1993) Nature 363: 418
101. Chong AKG, Pegg MS, Taylor NR, von Itzstein M (1992) Eur J Biochem 207: 225
102. Varghese JN, McKimm-Braschkin J, Caldwell JB, Korrt AA, Coman PM (1992) Proteins 14: 327
103. Goodford PJ (1985) J Med Chem 28: 849
104. Taylor GL (1993) Nature 363: 401
105. Varghese JN, Colman PM (1991) J Mol Biol 221: 473
106. Poirrette AR, Artymiuk PJ, Grindley HM, Rice DW and Willett P (1994) Protein Science 3: 1128–1130
107. Hurley JH, Dean AM, Sohl JL, Koshland DE Jr, Stroud RM (1990) Science 249: 1012
108. Bersohn M, Fujiwara S, Fujiwara Y (1986) J Comput Chem 7: 129
109. Heringa J, Argos P (1993) Proteins Struct Funct Genet 17: 391
110. Kasinos N, Lilley GA, Subbarao N, Hanneef I (1992) Prot Eng 5: 69
111. Koch I, Kaden F, Selbig J (1992) Prot: Stuct Funct Genet 12: 314
112. Rufino SD, Blundell TL (1994) J of Computer-Aided Molecular Design 8: 5
113. Rooman MJ, Wodak SJ (1988) Nature 445: 45

Identification of Structural Similarity
of Organic Molecules

Yoshimasa Takahashi

Department of Knowledge-Based Information Engineering, Toyohashi University of Technology,
1-1 Hibarigaoka, Tempaku, Toyohashi 441, Japan

Table of Contents

1 Introduction 106

2 Structural Feature Analysis 106

3 Identification of Common Structural Features 110
  3.1 Topological Common Structural Features 110
  3.2 Three-Dimensional Common Structural Features 115

4 Commonality to Similarity 122

5 Quantification of Structural Similarity 126

6 Conclusion 131

7 References 132

This review focuses on the structural aspects and discusses several approaches to computerized structural feature analysis and automatic identification of structural similarity of organic molecules with some illustrative examples relating to the structure–activity problems.

Topics in Current Chemistry, Vol. 174
© Springer-Verlag Berlin Heidelberg 1995

1 Introduction

It is well known that a chemical structure diagram or three-dimensional structure of a complex molecule contains information related to its biological activity, reactivity, and various other physicochemical properties. There are several possible ways of describing the structural feature of a molecule [1, 2, 3]. Binary descriptors, that is, the presence or absence of a particular substructural fragment, are widely used a screens in chemical structure search. The use of counts of the fragments have been also employed [4]. In the area of structure-activity studies and molecular design, the investigation of structural features such as functional groups or substructural fragments, common to a group of chemical compounds exhibiting a particular biological activity often leads to very useful and objective information [5]. In addition, 2D an 3D maximal common substructure search may provide some hypothetical pharmacophore [6]. Furthermore, for the intellectual use of computer in future chemical research it also will be quite important how to handle "similarity", especially molecular similarity or structural similarity on a computer [7].

This review discusses several approaches for the automatic identification of common structural features or structural similarity of organic molecules. The organization of the chapter is as follows. Section 2 gives an overview of the methods for structural feature analysis. Identification of common structural features is discussed in Sect. 3 with a few applications in structure-activity studies, which is subsequently followed by the identification of structural similarity in Sect. 4. The quantification of structural similarity is discussed in Sect. 5. The basic algorithms of these approaches and the relative software systems are also referred to with some illustrative examples.

2 Structural Feature Analysis

An understanding of the structural features of chemical compounds is extremely important in structure-property (or activity) studies, in general. There are a lot of additivity models for the physical properties of chemical compounds [8]. These models can be established by group contribution methods where the correlation between the substructural fragments constituting the molecules and their properties are examined. Generally, such works require some structural feature analysis of the molecules. The structural feature analysis could be done by manual approaches for a small set of molecules. However, such methods become quite tedious and time consuming for a large set of molecules. Most of the computerized approaches use a set of predefined substructural fragments such as substructure-search screens of chemical information systems.

Adamson et al. studied the structural features of chemical compounds in a large computer-based file [9]. The features are based on substructural fragments of their chemical structures. Several applications with an automated extraction scheme of such a substructural descriptor has been reported in structure-property and structure-activity problems [10–13]. Substructural descriptors have also been used for the comparison of structural similarity and the clustering of chemical compounds based on it [14–18]. However, the analysis of structural features of the compounds is a process necessary for the recognition of similarity.

Takahashi et al. [19] presented an exhaustive method of structural feature analysis based on the enumeration of all the possible substructures of a chemical structure. Figure 1 shows an example for the generation of all possible unique subgraphs from acesulfame (6-methyl-1,2,3-oxathiazin-4-one 2,2-dioxide) in the simple graph expression in which all atom types and all bond types are equivalent. In this case, the total number of unique subgraphs is thirty nine: they consist of the subgraphs with size one (single edge graph) to ten (skeletal graph of acesulfame). The number of subgraph of each size follows as: 1, 1, 2, 3, 4, 7, 7, 8, 5, and 1. The number of all possible subgraphs which involve isomorphic ones at each size is 269. Figure 1 shows the result with a simple graph expression. There are several ways to describe a structure diagram in a graph. Two other graph expressions (node weighted graph and node & edge weighted graph expression) have been employed in the analysis. The results obtained with these three graph expression modes are compared in Table 1. As shown in Table 1, the number of unique subgraphs increases in the order simple mode, node weighted mode and node & edge weighted mode. This is due to the fact that the highly weighted graph expression makes the number of possible unique subgraphs increase at each size of subgraph.

Carhart et al. [15] described a generalized structural descriptor called an "atom pair" which is defined in terms of the atomic environments of, and shortest path separation between, all pairs of atoms in the topological representation of a chemical structure. A similar descriptor has been suggested by Klopman [16]. More recently, Judson [20] described a more sophisticated approach to analyze structural feature of molecules for structure–activity studies.

On the other hand, ring substructures are also very important features for the automatic interpretation of chemical structures. There are many algorithms for ring perception within a chemical structure [21]. It is known that finding all possible rings within a chemical structure is a time consuming task. However, not all of them are always needed. The set of rings we need often depends on the problem we are trying to solve. An exhaustive review of ring perception algorithms has been reported by Downs et al. [22]. They also presented an algorithm to find the extended SSSR (Smallest Set of Smallest Rings [23]) and its application to generate ring screens for substructure search in large databases [24]. And an alternative algorithm for the systematic perception and extraction of possible ring substructures (SSSR, SSSR-dependent elementary rings, fused rings, spiro ring substructures, and so on) has been devised [25]. An example of

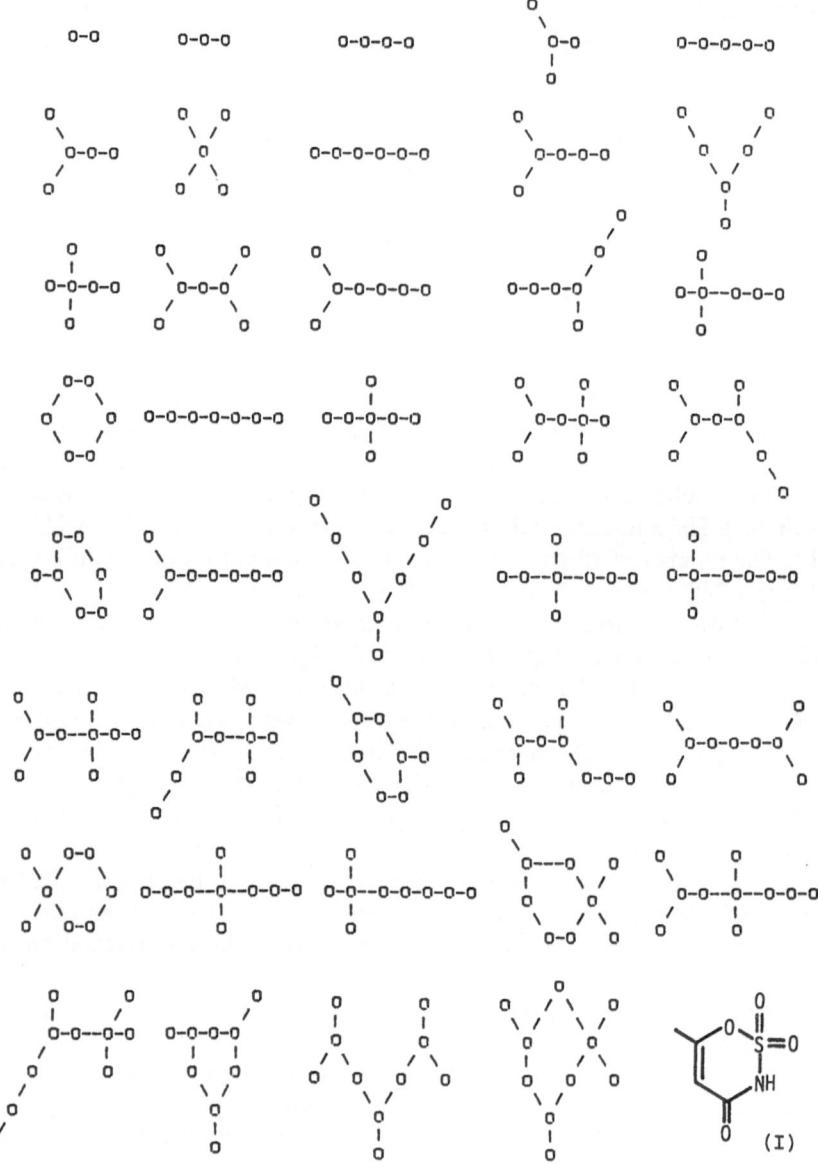

Fig. 1. All possible subgraphs of acesulfame (I) generated with the simple graph expression

the ring substructure analysis is shown in Fig. 2. It shows that they consist of six elementary rings (four SSSR members and two SSS-dependent elementary rings), ten bicyclic ring substructures, seven tricyclic ring substructures and a tetracyclic one that is the maximum one giving the full structure in this case.

Table 1. Computational results of the generation of all the possible unique subgraphs from three different graph expressions of acesulfame

	Simple graph	Node weighted graph	Node & edge weighted graph
Number of unique subgraph	39	170	207
All the possible subgraphs	269	269	269

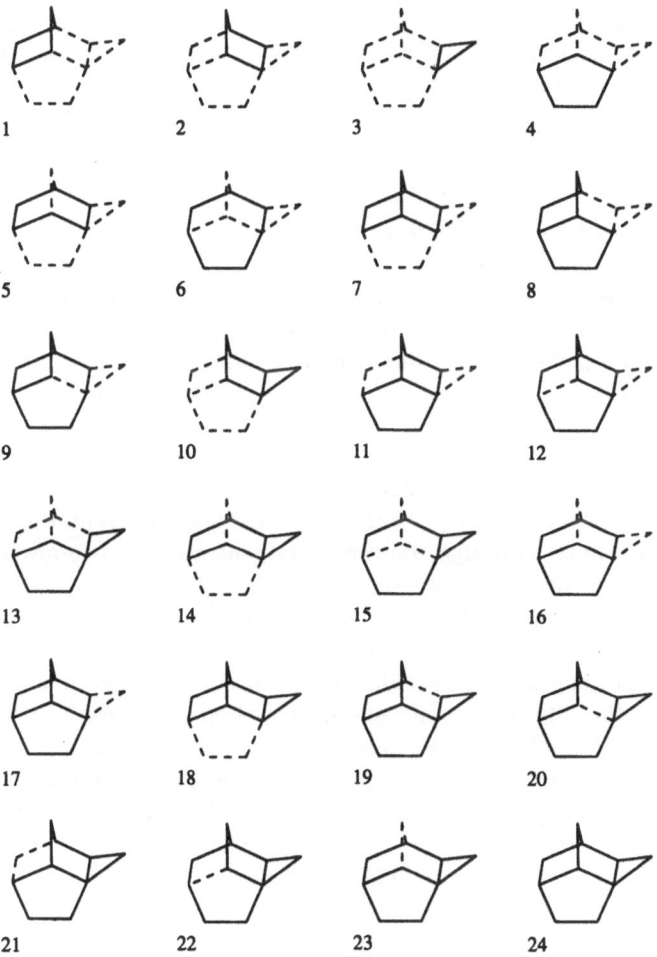

Fig. 2. All possible ring substructures of structure **24** in the figure. The first four simple rings (1–4) are part of the SSSR, and 5 and 6 are SSSR dependent elementary rings. These are monocyclic ring substructures. Rings 7–16 are bicyclics. Rings 17–23 are tricyclics. Ring 24 is tetracyclic, which is the full structure in this case

3 Identification of Common Structural Features

There have been several important attempts to determine structural fragments which are common among a group of chemical compounds. In these studies, structural features of compounds having identical or similar properties are compared in order to identify common structural fragments which would be closely related to their properties. In the field of medicinal chemistry, in particular, investigations of structure–activity relationships constitute an active area of research. For a group of compounds which show identical or similar biological activities though they are different in chemical structure, it is assumed in these studies that the activities in question should be attributed to a particular molecular feature which the compounds have in common. In this section, several approaches for the automatic identification of topological, 2D and 3D common structural features among organic molecules are discussed.

3.1 Topological Common Structural Features

The automated recognition of the largest connected common substructure has been attempted by several researchers [26–28]. The algorithm reported by Lynch et al. [26] was applied to analyze the structural similarities between two compounds. That by Cone [27] comprises dot plot notation [29] based on Wiswesser line notation [30] and provides an interesting approach based on mass spectral fragmentation. These two are pair wise approaches, which work exclusively in cases where just two chemical structures are considered. On the other hand, the algorithm developed by Varkony et al. [28] can deal with three or more structures at a time. This algorithm includes generation of two-atom structures in all of the given structures, and, through implementation of complete matching, the two-atom-structured graphs are allowed to grow. However, when large structures are encountered, the process requires not only a long processing time in the step of subgraph generation for all of structures time but also a large amount of memory for the data storage. In a modified Varkony's approach [31], selection is first made of a reference standard structure from the number of chemical structures given. Subsequently, two-atom fragmental structures are generated, and the resultant graph is developed to growth. The determination of the common substructural fragments generated during the growth of the graph is performed by means of a substructure searching technique.

Commonality should be strictly defined when common substructures are searched for. The following five levels of the representation of a common feature may be defined from a graph theoretical point of view (Fig. 3).

(i) Simple graph representation, where difference in atomic species in the molecular structure, as well as the multiplicities of the chemical bonds, are all

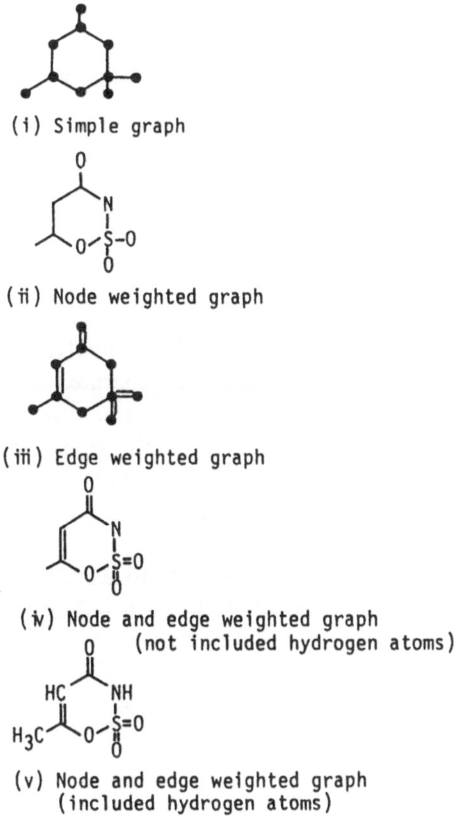

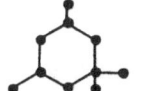

(i) Simple graph

(ii) Node weighted graph

(iii) Edge weighted graph

(iv) Node and edge weighted graph
(not included hydrogen atoms)

(v) Node and edge weighted graph
(included hydrogen atoms)

Fig. 3. Various graph expressions of the chemical structures of acesulfame

neglected, and all the atoms are regarded as equivalent. This is a skeletonized expression of structures.

(ii) This is a node-weighted representation, where the atomic species is recognized but the multiplicity of the bonds is neglected.

(iii) This is an edge-weighted representation, a reverse of level (2).

(iv) Node- and edge-weighted representation. Different atoms and bonds are all differentiated in recognition. But, all hydrogen atoms are neglected in the description of a chemical structure. This is an expression from a chemical point of view.

(v) In addition to level (iv), hydrogen atoms are also incorporated in structural representation. Therefore, this provides a more precise chemical expression than the preceding level.

In this chapter, the largest common substructure is defined as a connected subgraph having the greatest number of edges in a graph. The algorithm of Takahashi et al. [31] is summarized as follows,

Step (i): A reference standard structure (as a graph) is selected out of the given plural number of structures (graphs).

Step (ii): The reference standard graph is degraded into a set of subgraphs with size 1.

Step (iii): Each of the resultant subgraphs are made canonical then any duplicated representations are checked.

Step (iv): Detection of common subgraphs in the given chemical structures can be done by the subgraph (substructure) search technique. If there are no common subgraphs present, the search is terminated.

Step (v): The common subgraphs obtained hitherto are stored in the candidate list.

Step (vi): The subgraphs stored in the list are allowed to grow.

Here, the size of the subgraph means the number of edges which are involved in the subgraph. All the possible subgraphs larger by one are generated and permitted to grow to their maximal limit (and the flow of logics returns to Step (iii)). If a greater subgraph cannot be generated any longer, the algorithm comes to an end. Then, the substructural fragment corresponding to the subgraph stored in the candidate list gives the greatest and common substructure.

Figure 4 illustrates the largest substructural fragment common to acesulfame and saccharin, both being artificial sweeteners, extracted on the basis of level (1)

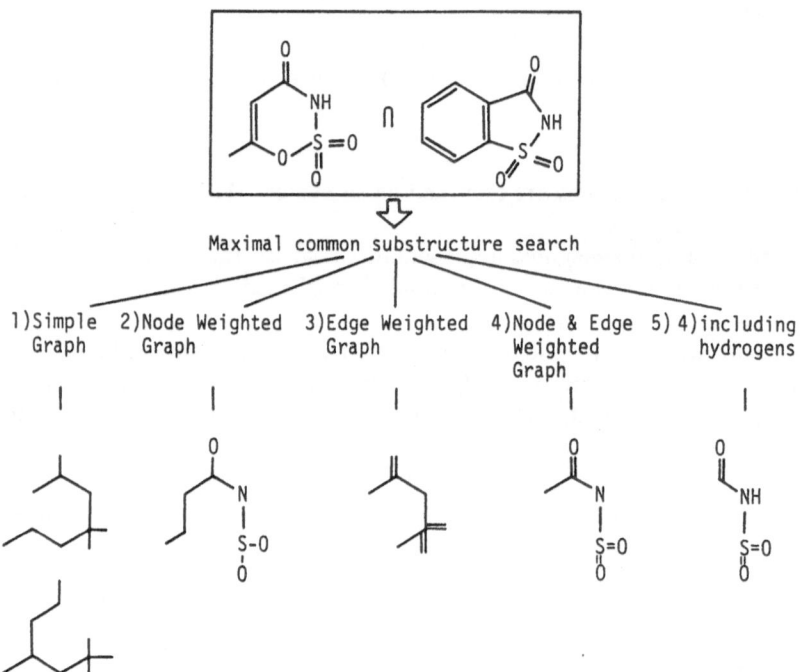

Fig. 4. Largest common substructures founded by MAXFIT [31] with different levels of commonalities

to level (5) criteria. At level (1), two connected graphs, each consisting of 9 edges, are recognized. From Level 2 on, as the weights are accumulated, the size of the largest common substructure (the number of edges) becomes smaller. Table 2 summarizes the results with the number of all possible common substructures at each level. Among these, all the 24 different common substructures conceivable at level (4) are shown in Fig. 5.

Table 2. Results of the common substructure analysis between acesulfame and succharin

Search level	Maximal size	The number of unique common subs.
1	9	32 (254)[a]
2	8	44 (91)
3	7	25 (97)
4	6	24 (31)
5	5	15 (18)

[a] The number in parentheses is the total number of common substructures which involve isomorphic ones

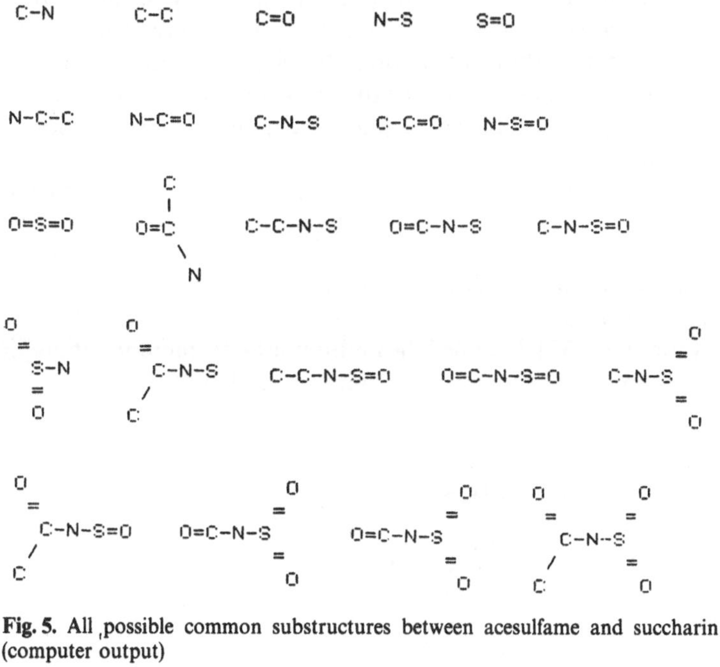

C-N   C-C   C=O   N-S   S=O

N-C-C   N-C=O   C-N-S   C-C=O   N-S=O

$$O=S=O \quad O=C\overset{\displaystyle C}{\underset{\displaystyle N}{|}} \quad C-C-N-S \quad O=C-N-S \quad C-N-S=O$$

$$\overset{O}{\underset{O}{\overset{\|}{S}}}-N \quad \overset{O}{\underset{C}{\overset{\|}{C}}}-N-S \quad C-C-N-S=O \quad O=C-N-S=O \quad \overset{}{C}-N-S\overset{O}{\underset{O}{\overset{\|}{}}}$$

$$\overset{O}{\underset{C}{\overset{\|}{}}}-N-S=O \quad O=C-N-S\overset{O}{\underset{O}{\|}} \quad O=C-N-S\overset{O}{\underset{O}{\|}} \quad \overset{O}{\underset{C}{}}C-N-S\overset{O}{\underset{O}{\|}}$$

Fig. 5. All possible common substructures between acesulfame and succharin with the level 4 (computer output)

Another example of the largest common substructure search is shown in Fig. 6. Although they differ greatly in structure from one another, morphine, meperidine, and methadone exhibit similar pharmacological activities (analgesic properties). By the search with level (4) expression, two connected graphs are

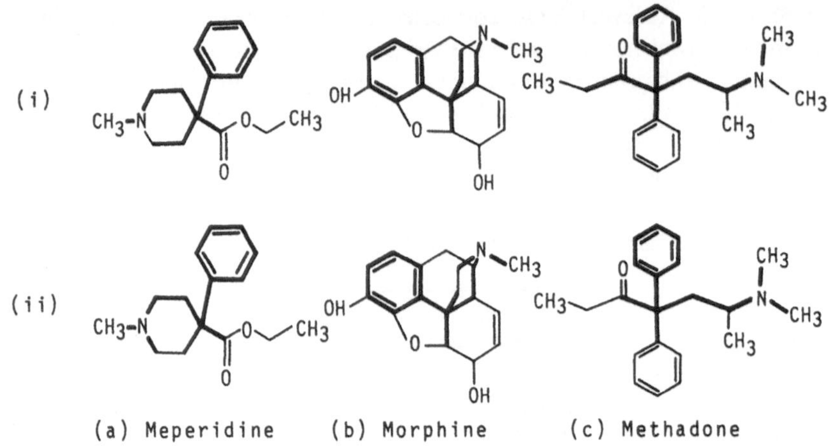

(i)

(ii)

(a) Meperidine   (b) Morphine   (c) Methadone

Fig. 6. Largest common substructures found for three narcotic analgesics under level 4

obtained as shown in Fig. 6. It has long been pointed out that the simultaneous presence of a phenyl group and a nitrogen-containing site in the molecular structure is a prerequisite for a compound to act analgesically on the central nervous system. The largest substructures common to these three compounds were found in the present study to be a connected graph including a phenyl group and a nitrogen-containing site, it was then possible to conclude that the substructures are the "topological pharmacophore" required in the expression of analgesic activities in these compounds.

It is interesting that in the 3-dimensional structures of them, obtained from X-ray diffraction data [24–26], they can be superimposed on one another at the point of the largest common substructure [31]. Figure 7 shows a three dimensional stereoscopic view of the common part of them, which were superimposed by the least squares method.

Recently, Bayada et al. [32] described the multiple largest common subgraph problem mentioned above with an efficient algorithm to handle it.

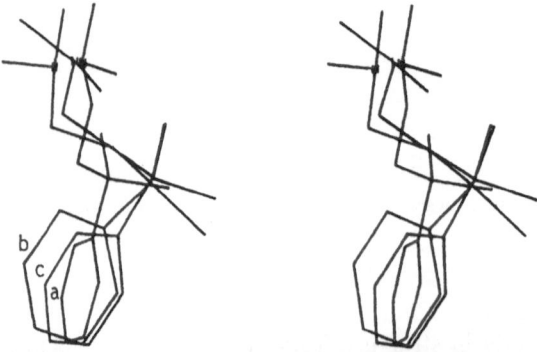

Fig. 7. A stereoscopic view of the three-dimensional superimposition of the largest common substructure, (ii) in Fig. 6, among three narcotic analgesics

3.2 Three-Dimensional Common Structural Features

The topological approaches mentioned above are quite useful for the structural feature analysis for a large set of molecules [5]. For biologically active compounds, however, structural factors which contribute to the development of pharmacological effects should be investigated, not only topologically, but also on a three-dimensional level in order to permit a more detailed analysis of the relationships between structures and activities [33]. Recently, computer graphics has become more widely used and techniques for utilizing it have progressed, making it possible to look through several structures that overlap each other [33–36]. another interesting technique is called the "three-dimensional pharmacophoric pattern search or 3D substructure search" [33, 37–43]. A query structure is designated at a three-dimensional level, in a similar manner to the conventional substructure search, and an operation is specified to establish if it exists simultaneously in two or more structures. Other attempts [44–48] have also been made to examine structures overlapping each other and to compare the molecular shapes quantitatively.

In addition to investigations by mere structure overlapping methods and retrieval of three-dimensional substructures, there have been studies focused on developing an automated operation for recognizing common substructures, which takes into account the three-dimensional geometry of the molecules [49–54]. These approaches are greatly different from conventional one in that the designation of a query structure is not required.

The COMPASS (COMmon geometrical PAttern Search System) algorithm [54] is described below and was designed to search for a pattern of common three dimensional geometry. The basic idea of the algorithm is based on the distance geometry approach proposed by Crippen et al. [55, 56]. First, three-dimensional coordinate data of each molecule are given as the input. If all atoms are assumed to be equivalent, the structure of each molecule can be expressed by a set of points in the three-dimensional rectangular coordinate space. Then, it becomes possible to define the geometry of each molecule by simply using a distance matrix corresponding to the set of points. Such distance matrices are easily developed from three-dimensional atomic coordinate data of the molecules. A distance matrix so produced is considered to represent an edge-weighted complete graph which consists of the same number of nodes as that of constituent atoms in the molecule. Thus, the graph-theory approach can be applied to the search for a three-dimensional common geometrical pattern. The complete graphs for the two molecules are compared with each other according to the criteria of commonalities based on edge-weight, and then a new graph can be produced called a docking graph [57]. The docking graph formed is either a connected or a disconnected graph and contains information on the edges with the same weights (under prescribed conditions such as allowance for distance) as those given to the above-mentioned two complete graphs (Fig. 8).

Therefore, searching for some edge-weighted maximal common subgraph for the two complete graphs is equivalent to searching for a clique [58] in this

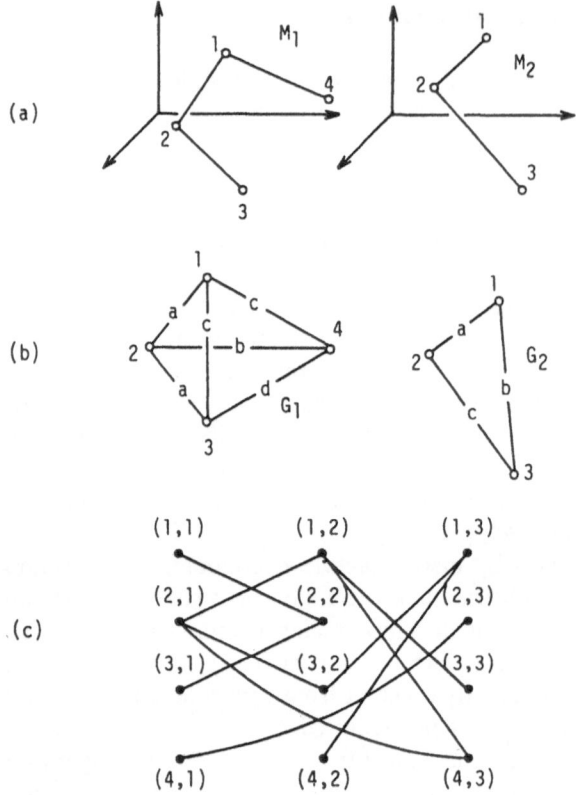

(a)

(b)

(c)

Fig. 8. Docking graph from graphs G1 and G2

docking graph. The set of nodes of each subgraph, which is a candidate for the clique, obtained in this search process corresponds to the set of constituent atoms of the desired common geometrical pattern. And the set of nodes consisting of the clique finally obtained is correlated with the constitution of the maximal common geometrical pattern and the set of their atoms.

A major problem with expressing a three-dimensional structure by a distance matrix is that the geometry of enantiomers cannot be distinguished. No information on enantiomer geometry is incorporated in the common geometric patterns which are obtained by the clique search method using a docking graph based on a distance matrix. For checking that geometry, three-dimensional coordinate data on the molecule are used in addition to the distance matrix in conducting the following operations: (1) combinations of four points ($_nC_4$, n denotes the number of nodes) are produced from the set of nodes which are contained in the common geometric pattern recognized on the basis of a reference molecule; (2) three of the vertices of each tetrahedron are selected arbitrarily to establish a reference plane, and the remaining one is taken as the probe point; (3) the reference plane is then looked at from such an angle that the constituent nodes with increasing code numbers are clockwise. The tetrahedral geometry is represented by " + 1" if the probe point is beyond the plane, by " − 1" in the contrary case, and by "0" if all four points are in the same plane or can be regarded as being in the plane in view of the tolerance (Fig. 9).

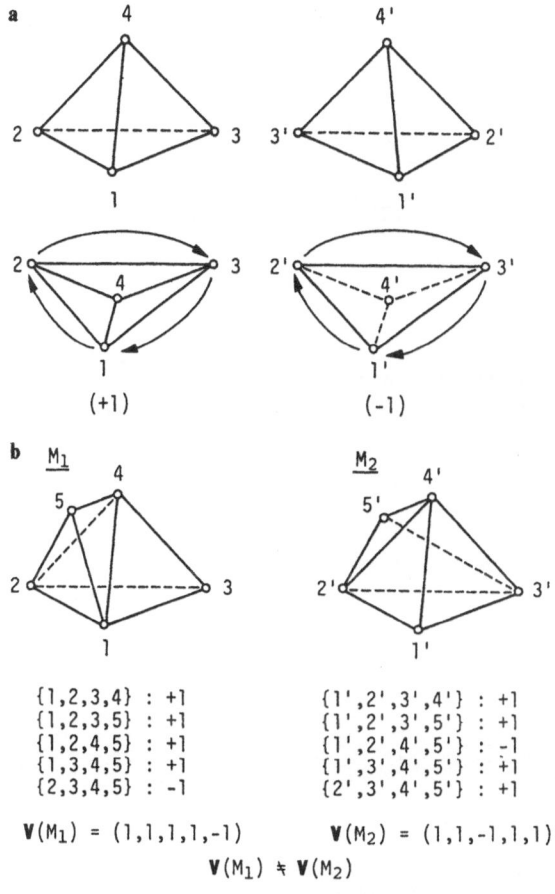

Fig. 9. Designation of configuration pattern vector in polyhedral geometry: **a** tetrahedral reference configuration pattern and the corresponding sample configurational pattern; **b** Generation and comparison of the configurational pattern vectors for two enantiomer polyhedral geometric patterns

In this way, a spatial pattern vector is developed which can represent the configurations of all possible tetrahedral geometries resulting from the geometric pattern in question, by three values (i.e. 1, − 1 or 0). Such a pattern vector for the geometric pattern based on the other molecule to be compared is generated with the reference plane corresponding to that of reference molecule. The enantiomer geometries of two distant-isomeric patterns can be distinguished from each other by comparing these two configurational pattern vectors.

```
EXECUTION =========================
    INPUT-FILE :IN
    OUTPUT-FILE:OUT
    DATASET    :DATASET
-- SAMPLES --
    SAMPLE(1)= F02(Haloperidide)  (32)
    SAMPLE(2)= F10(Milenperone)   (29)
-- MATRIX SELECTION --
    DISTANCE MATRIX
    ALLOWANCE=  .50
-- MUTUAL CHECK --
    OFF
-- STEREO CHECK --
    OFF
-- NODE WEIGHT --
    ATOM TYPE=ON
    BOND TYPE=OFF
    CHARGE TYPE=OFF
-- NODE EXCLUSION --
    EXCLUSION    = NOT
-- FUNC. GROUP PRESERVATION --
    PRESERVATION = NOT
-- FUNC. GROUP COMPRESSION --
    COMPRESSION  = NOT
-- RESULT --
  ROTATION    1/   0
NO.  1
 (F02: 1)    2   3   4   5   6   7  20  21  23  29  31  32
 (F10: 1)    3   2  11  10   9   4  12  20  18  26  23  27
      2)     3   2  11  10   9   4  12  20  18  26  29  27
NO.  2
 (F02: 1)    2   3   4   5   6   7  20  23  29  30  31  32
 (F10: 1)    3   2  11  10   9   4  12  18  26  28  29  27
NO.  3
 (F02: 1)    2   6   7   8  17  18  19  20  21  22  26  30
 (F10: 1)   24  21  23  18  16  15  20  19  14  13  10   2
NO.  4
 (F02: 1)    2   6   8  16  17  18  19  20  21  22  26  30
 (F10: 1)   24  21  18  17  16  15  20  19  14  13  10   2

              NUMBER= 4
              SIZE  =12
```

The COMPASS serves to automatically determine common geometrical patterns among three-dimensional molecular structures. Three-dimensional coordinate information of the molecules is used for the calculation. In addition, such data as the charge and connection table of each atom are also input when required. The latter data are utilized for weighting the atoms to express their atomic species, environmental conditions and other characteristic features. Charges of atoms can be assumed within a certain tolerance. By this processing operation, differences in atom type and electronic environment, as well as interatomic distances, can be reflected in the process for generating the docking graph to be used as the basis of the search for common geometrical patterns. This operation, therefore, makes it possible to search for three-dimensional common geometric patterns which have distinctive chemical meanings (such as those called pharmacophores). Furthermore, it is also possible to search for maximum common geometrical patterns that contain some partial structures or geometrical patterns specified by the user.

An example for the search of three-dimesional maximal common geometrical patterns between two molecules (haloperidide and milenperone) by COMPASS is presented in the following. These molecules are known as dopamine antagonists. These atomic coordinates were taken from references [59, 60]. Their chemical structures and results of the COMPASS operation are shown in the form of computer output in Fig. 10. For the search conditions, a distance tolerance of 0.5 Å was used and only the atomic species is used as the weighting parameter. Four maximal common geometrical patterns with 12 constituent atoms were obtained as given in Fig. 10, in which the numbers following the code numbers (F02, F10) indicate the numbers assigned to each atom in the structural formulae. In case No. 1, two common geometric patterns for milenperone (F10) with one position occupied by a different constituent atom (23 or 29) are obtained as output, corresponding to a geometrical pattern with 14 atoms, ie. 2, 3, ... , 32, which is contained in haloperidide (F20). This resulted from the use of distance tolerance only. These results can easily be visualized by computer graphics for displaying superimposed structures. Figure 11 shows structure superposition based on interatomic least square fitting for the first pattern in case No. 1 in Fig. 10.

Another search trial [54] for the above two molecules was done by using additional information on the charges of the atoms in each molecule, "charge" in this case means the net charge obtained by a semiempirical molecular orbital calculation. Two modes of charge designation can be used in the system.

In one, a tolerance is designated in terms of the relative difference in charge between particular atoms to be compared (based on an idea similar to that for distance tolerance); in the other, the weighting operation is performed automatically according to range designation. The result in Fig. 12 was executed with

---

Fig. 10. Maximal common geometric patterns found by COMPASS [54] for haloperidide and milenperone. In weighting atoms, only different atomic species were considered for this search. The distance allowance is 0.5 Å

Y. Takahashi

SKLTN BA&ST SPLIN SPDOT DOTSF HATCH SINGL STEREO QUIT

CHARG ESP

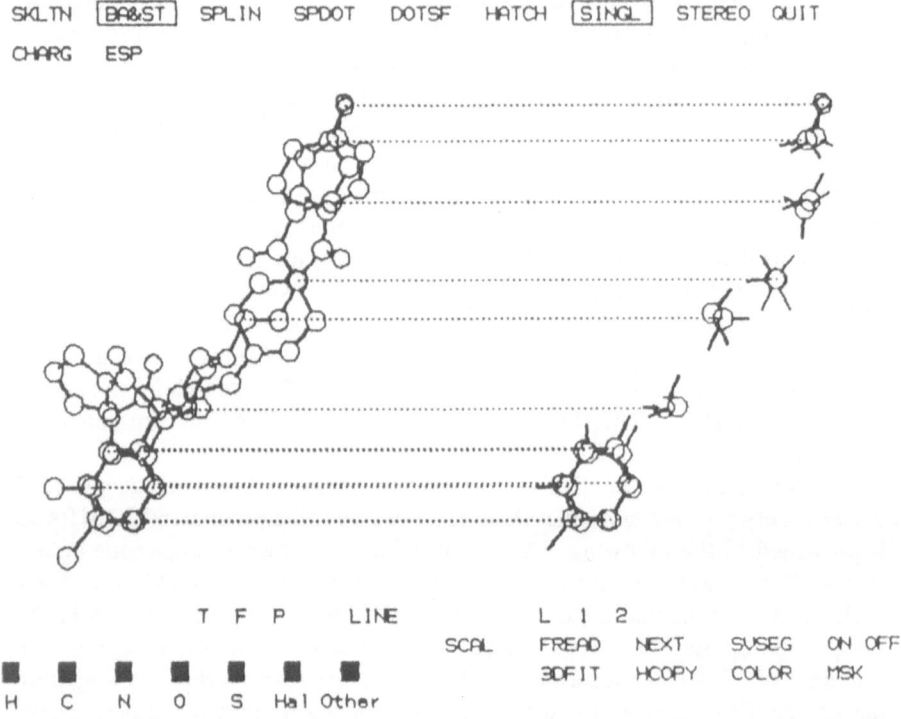

T F P LINE   L 1 2

          SCAL   FREAD   NEXT   SVSEG   ON OFF

          3DFIT   HCOPY   COLOR   MSK

H C N O S Hal Other

Fig. 11. Graphic representation for the first results in Fig. 5. The *left-hand view* shows the superimposition for the whole molecules, except for hydrogen atoms, and the *right-hand view* is for the relevant portions

the latter mode and the atoms were weighted in three steps, i.e., $N \leq -0.100$, $-0.100 < N < 0.100$ and $0.100 \leq N$. Other search conditions including distance tolerance are the same as those used in the previous case. COMPASS finally provided eight maximal common geometric patterns, each composed of eleven constituent atoms. Figure 12 displays the most understandable result of those obtained. The fluorine atom in the *p*-substituted benzoyl, the other benzene rings and carbonyl oxygen in the imido group in each molecule correspond consistently to their counterparts in the other molecule. Not only do the common portions of the two molecules displayed in Fig. 12 agree well with each other in terms of three dimensional geometry, but also each atom in one molecule, as expected from the search conditions, has a similar electronic environment to that of its counterpart in the other.

The above examples used only a pair of molecules to illustrate the approach. To discuss three-dimensional structure-activity relationships in more detail from a strict pharmacological point of view, common pattern search trails should be conducted for a large number of molecules which have similar activities and different structures. In practice, COMPASS can work for three or more

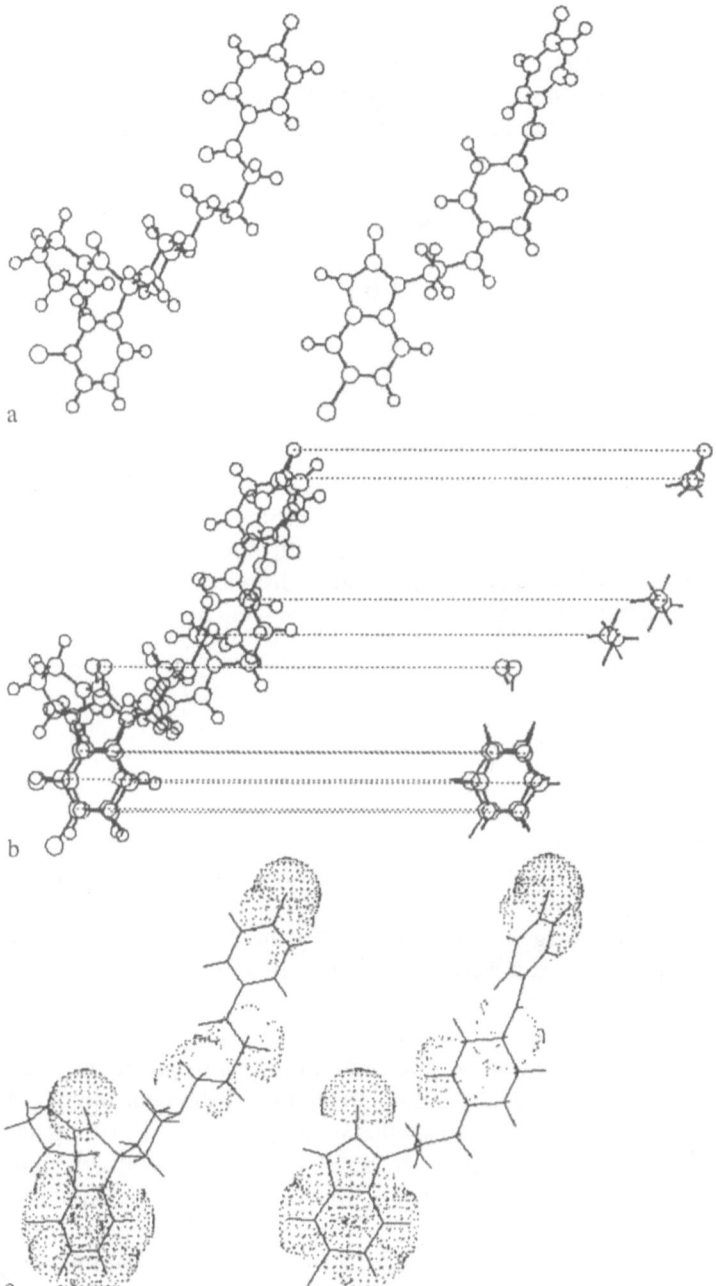

Fig. 12. The maximal common geometric pattern founded by COMPASS. In weighting atoms, different atomic species and atomic charge were considered. The distance allowance is 0.5 Å. **a** Three-dimensional structures of haloperidide (*left*) and milenperone (*right*). **b** Display of the superimposition of their whole molecules and common parts. **c** Display of the common parts with electronic property

molecules [54]. To achieve more practical analysis of structure–activity relationships, however, the conformation of each molecule should be examined thoroughly prior to the common geometrical pattern search, since possible conformations of a molecule under some assumption for actual interaction with a receptor do not always agree with those determined from X-ray diffraction data. If there are several conformers which appear to be stable in terms of energy, each of them should be examined in conducting the analysis. Once reasonable three-dimensional molecular coordinates are obtained during such a process, this kind of system works as a powerful tool to provide candidate patterns which may be closely related to the biological activities of the molecules.

4 Commonality to Similarity

On the problems related to structure–activity relationships it is often pointed out the relevance of structural features associated with some of substructures mutually apart from each other with particular distance [5, 15, 61]. Now consider the two structures in Fig. 13. "Are these two similar or not?"

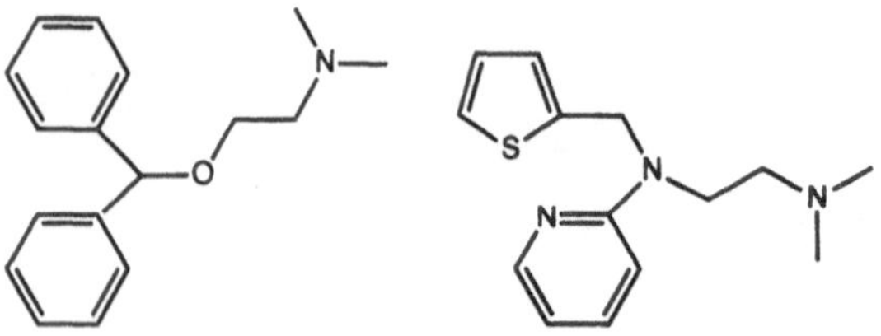

Fig. 13. Are these two similar or not?

If one focuses on the largest common substructure, one can just get a dimethylaminoethyl moiety. So, it might be said they are not very similar. From the other point of view, however, one may say they are very similar because both of them have two aromatic rings and a tertiary nitrogen atom, in addition the topological path length between one of the rings and the nitrogen is the same with 5 in both cases. In actual practice, both of them possess the same biological activity, i.e. the antihistamine activity. From this stand point, a novel method has been devised for locating the "common" or "similar" structural features composed of several substructures corresponding to that just mentioned above [62].

This approach uses, as the sole input, the structural information in the connection table for the chemical compound. However this information is not handled directly in that form, but is transformed into a reduced structural representation that is based on functional groups. This representation is an abstracted graph formed by the connection of all the nodes after they have been corresponded to the functional atomic groups. Every edge of the graph is weighted by the topological path length between the atomic groups related to the terminal nodes of each edge. Accordingly, the reduced representation of chemical structure simplifies the exclusion of parts of the molecule that are not relevant in a structure activity relationship analysis, replacing them by means of a 'spacer', and thus increasing the effectiveness of the structure handling process. Outline of the generation of the reduced representation is as follows.

First, the system reads the connection table and proceeds with the pre-processing of this information, registering the node numbers of the functional groups perceived in the structure. The functional group perception can be performed by means of a substructure search technique. A molecular structure, however, frequently possesses many interrelations between atoms each of them yielding different functional groups. When this happens, the system assumes all the possible combinations and looks for all the functional groups that are in agreement with the desired ones. The functional atomic groups to be considered are defined in advance and stored in a special file (substructure definition file) by the user.

Once all the functional atomic groups are perceived, the interrelations between them are checked. The relationships evaluated are described in terms of matrix expression and they are divided into two cases:

i) The case in which two functional groups are partially overlapping.
ii) The case in which one of the functional atomic groups is completely included by another one.

For the former case, the relationship is described in the overlapping matrix, and the latter is described in the inclusion matrix to avoid the duplication of nodes in the reduced graph representation. They are used later at the stage of determination of path length between functional groups and at the stage of docking graph construction.

In the next step, the distance between functional atomic groups is determined. This distance is called the topological path length. If the structure has ring(s) there are several possible paths that can be drawn simultaneously between functional groups. In such a case, some of edges are weighted with multiple values. Consequently, the chemical structure is expressed as a graph with multiple edge weights. Figure 14 exemplifies such an expression of a chemical structure.

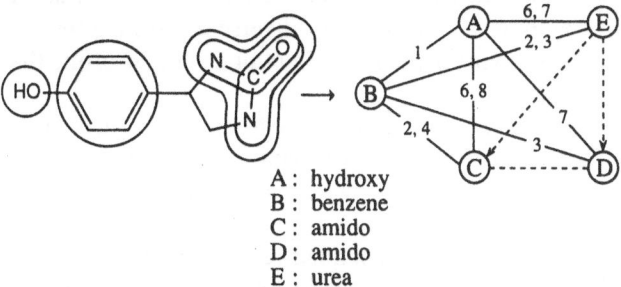

A : hydroxy
B : benzene
C : amido
D : amido
E : urea

Fig. 14. Example of the reduced-graph expression of a chemical structure

The reduced graph representation in Fig. 14 corresponds to a graph composed of weighted edges and with nodes equal to the number of functional atomic groups perceived. Thus, the problem can be handled by means of a graph theoretical procedure. The process of common feature search between graphs is based on the clique findings algorithm that were referred to in the previous section. It consists of two basic steps: 1) the generation of the docking graph of the two graphs, and 2) the determination of the maximal cliques of the docking graph. The docking graph, here, is an attempt to map pairwise the identical functional groups in the two original graphs. the mapping works if the distances between any two groups also fit. A clique in the docking graph corresponds to a grouping of functional groups in the original graphs, where all the intragrouping distances are the same in both of the original graphs.

Now, returning to the structures in Fig. 13, one can still get a dimethylaminoethyl moiety as the largest common substructure between the two. Now the problem is how to obtain the "similar" structural features such as a topological pharmacophoric pattern. It can be overcome by setting up equivalence groups of edge pairs and/or node pairs in the process of docking graph generation. Allowances for differences in edge weights can be specified when corresponding edges during docking graph formation. On the other hand, a node isosterism definition file, which is a kind of substructure knowledge file, can take into account extended commonality (i.e., similarity) for the nodes.

The contents of the substructure knowledge file definitely depends on the problem at hand. In this work, knowledge on the functional isosterism was defined and stored as substructure knowledge from the stand point of structure–activity relationships. However, functional atomic groups may be correspondingly based on a variety of different rules. In this way, the system can refer to the functional groups from several different points of view. For example, –NH– group may be referred to by several means (a member of amine fragments, H–bonding donor, or electron pair donor). Thus, even if the real fragment does not survive in a "common" structural feature, the alternative feature may be concluded as the "similar" structural feature under the reasoning of the substructure knowledge. Figure 15 illustrates such the substructure knowledge file.

Two brief examples are presented to illustrate structure–activity problems for which this approach might be useful. The first example (Fig. 16) finds the similar structural features among five structurally diverse antihistamines (diphenhydramine, methapyrilene, cyproheptadine, dimethindene and promethazine).

In this case, all of the structures possess at least two aromatic rings and a tertiary nitrogen atom. Furthermore, their two aromatic rings and a tertiary nitrogen atom are separated from each other by four to six bonds respectively, and the two aromatic rings are separated by either two or three bonds. In promethazine, two nitrogen atoms are corresponded to the common feature of the nitrogen atomic group. Because of that, all possible paths were examined for the distance determination and edge weighting when generating the reduced

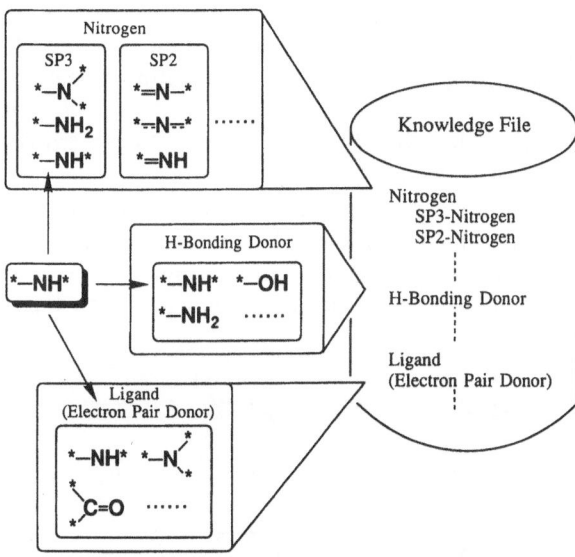

Fig. 15. Illustrative scheme of the substructure knowledge file

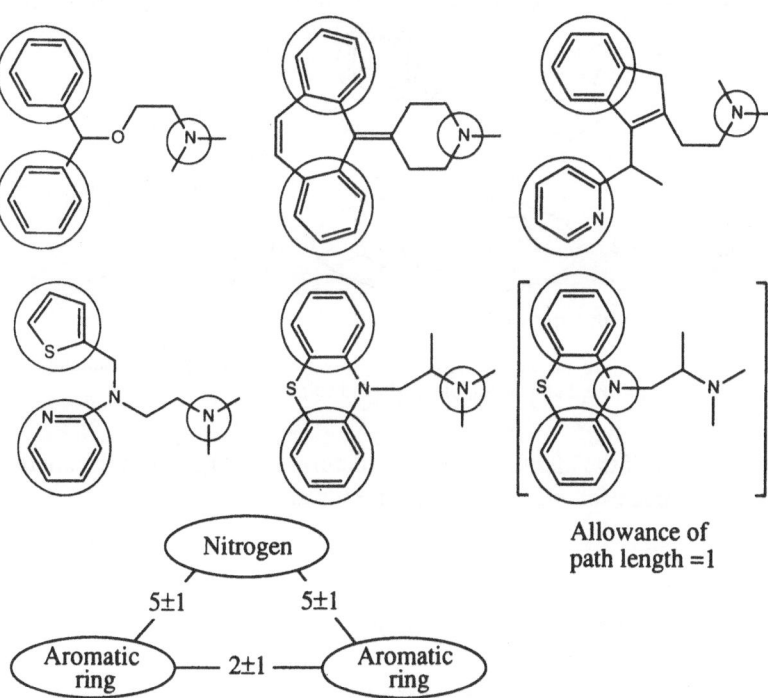

Fig. 16. Similar structural feature among five antihistamines (diphenhydramine, methapyrilene, cyproheptadine, dimethindene and promethazine)

graph representation of the chemical structure. The result is that these five antihistamines have the structural similarity that is expressed in the topological triangle based on the two aromatic rings and a tertiary nitrogen atom as shown in lower part of Fig. 16.

As another illustration, a structurally more diverse set of six antipsycotropic agents (molindone, chlorpromazine, chlorptothixine, haloperidol, pimozide, and sulpiride) were selected for the search of their molecular similarities. The result is summarized in Fig. 17. The result obtained here shows that all of these structures have a tertiary nitrogen atom, an aromatic ring and another aromatic ring (or a carbonyl group) that might be related to the π-electron interaction with their receptor.

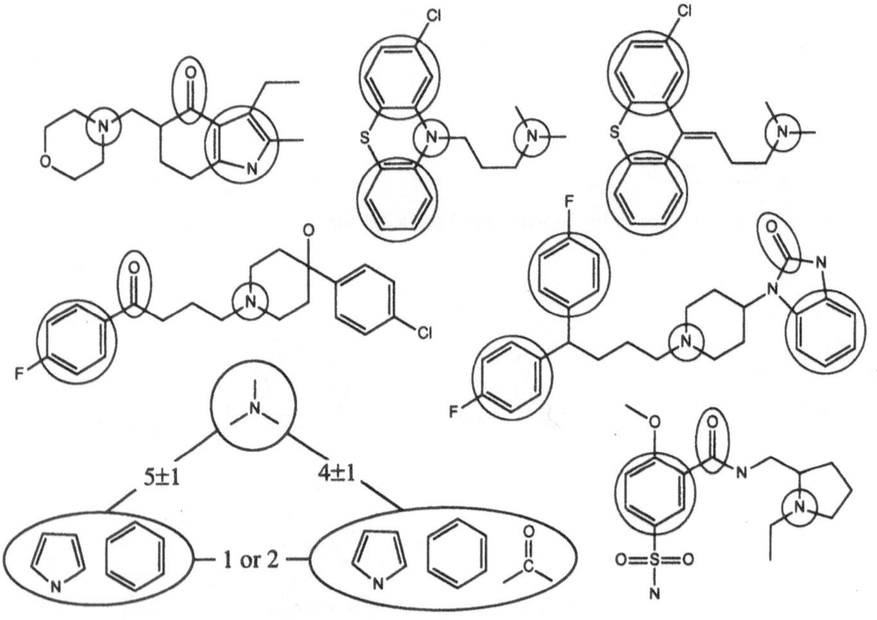

Fig. 17. Similar structural feature among six structurally diverse antipsycotropics (molindone, Chloropromazine, chlorptothixine, haloperidol, pimozide and sulpiride)

The basic concept of the identification of structural similarity discussed here could be easily extended for 3D molecular structures.

5 Quantification of Structural Similarity

In the preceding two sections, several approaches for the identification of common structural features, and of structural similarity have been presented. It should be noted that there is another aspect of the similarity concept. That is, how similar are the objects. In the chemical context, some quantitative criterion is generally available to initiate such an evaluation. Several methods and algorithms have been designed for implementation on a computer. Most of them were developed in structure–activity and structure–property studies. Substructural analysis method described by Cramer [64] and Adamson and Bush [10] gives a basis for approaching this problem. It is based on the correlation of activities or properties of chemical compounds with the occurrence of particular partial structures which are embedded in their structures. Adamson and Bush [14] also considered quantification of the similarity between chemical structures on the basis of the substructural analysis. In their paper, 39 compounds with local anesthetic activity were classified numerically by calculating similarity or dissimilarity coefficients between pairs of the structures and applying cluster analysis to the results. On the basis of this approach, Willett and Wintermann [64, 65] examined the effectiveness of various measures of the evaluation of molecular similarity for the clustering of chemical compounds. These approaches are based on the generation of substructural descriptors from chemical structures and their use in quantitative evaluation by any similarity measure.

The graph theoretical analysis has also been applied to the ordering of chemical structures and subsequent comparison of their relative properties [66–68]. A topological index is a numerical value and it represents a topological nature of molecular structure which is often used for quantitative correlations with physical, chemical and biological properties [8]. The first structural index of topological nature was introduced by Wiener [69] to establish a quantitative relationships between the molecular structures and boiling points of saturated hydrocarbons. Hosoya [70] has proposed a topological index, called Z, to characterize the topological nature of structural isomers of saturated carbons. He has also pointed out that the index Z correlates well with the boiling points. On the other hand, Randic and Wilkins [71–72] analyzed the structural similarity or dissimilarity by using similar numerical indices to provide the ranking and ordering of molecular structures. A variety of indices and their applications have been reported with structure–activity studies [73, 74].

Recently we have presented a method of quantitative evaluation of structural similarity of chemical compounds that uses only topological structural information [75]. The method is based on the enumeration and characterization of all the possible substructures. The remaining part of this section describes the basic concept of our method:

Enumeration and characterization of the substructures. For a given structure represented as a chemical graph (hydrogen suppressed graph), all the possible substructures embedded in it are enumerated. Here, all the substructures of the parent structure (the original chemical graph) possessing different elements but represented by the same subgraph are taken into account. Subsequently, all the individual substructures are characterized quantitatively. To perform this characterization two methods can be used as follows: (1) The overall sum of degrees of the nodes composing each subgraph. (2) The overall sum of the mass numbers of the atoms (atomic groups) represented by the nodes of the subgraph. With the first method the chemical structure is represented by a simple connected graph where the hydrogen atoms are omitted, thus, the characterization of the structure depends only on the topology of the structural skeleton. In the second method attached hydrogen atoms are taken into account as augmented atoms and are represented by weighting correspondingly their respective graph nodes.

Fragment spectrum. A "topological fragment spectrum" (fragment spectrum in what follows) is defined with the histogram resulting from displaying the frequency distribution of a set of individually characterized substructures (structural fragments) according to the value of their characterization index. The fragment spectrum generated according to this definition is a representation of the structural profile of the chemical compound in question. Figure 18 is an example of the fragment spectrum for 2-Methylbutane when each substructure characterized by the sum of degrees of the graph nodes.

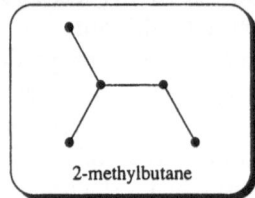

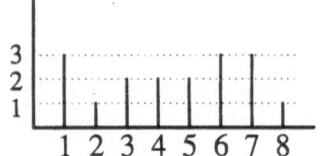

Fig. 18. A fragment spectrum of 2-methylbutane that was weighted by the sum of degree of the subgraph nodes

X=(3, 1, 2, 2, 2, 3, 3, 1)

Quantitative evaluation of molecular similarity. The fragment spectrum obtained in the above can be described as a kind of multidimensional pattern vector. Consequently, using this pattern representation of a spectrum it is possible to apply diverse quantitative methods for the evaluation of similarity.

Several types of similarity measures can be used to compare the similarity between such objects [4]. Two types of the most often used measures are distance-based similarity coefficients and correlation coefficients. Representatives of them are as follows:

(i) Similarity (S) based on the normalized distance:

$$S(X_i, X_j) = 1 - D(X_i, X_j)/D_{max} \tag{1}$$

X_i and X_j are the fragment spectra for the ith and jth molecule respectively. $D(X_i, X_j)$ is the Euclidian distance between the patterns X_i and X_j. D_{max} is the largest distance among the set of molecules under analysis.

(ii) Correlation coefficient (R):

$$R(X_i, X_j) = \frac{\sum\limits_{k=1}^{p} (x_{ik} - \bar{x}_i)(x_{jk} - \bar{x}_j)}{\sqrt{\sum\limits_{k=1}^{p} (x_{ik} - \bar{x}_i)^2 \sum\limits_{k=1}^{p} (x_{jk} - \bar{x}_j)^2}} \tag{2}$$

x_{ik} and x_{jk} are the kth quantitative characteristic index of the patterns X_i and X_j respectively. $\bar{x}_i$ and $\bar{x}_j$ are the averages of the quantitative characteristic indices and p is the number expressing the dimension of the pattern vector.

Figure 19 illustrates the spanning tree obtained by comparing the fragment spectra of twelve isomers of hydrocarbons constituted by five carbon atoms (three chains, five monocyclics, and four bicyclics), where the evaluation of similarity were carried out with Eq. (1); and the fragment spectra were obtained using the sum of the degree of the atoms at each vertex of the subgraphs. As is evident from the figure, the set of molecules is successfully grouped into chains, monocyclics and bicyclics. This result shows that the fragment spectrum successfully reflects the topological constitution of the molecular skeleton.

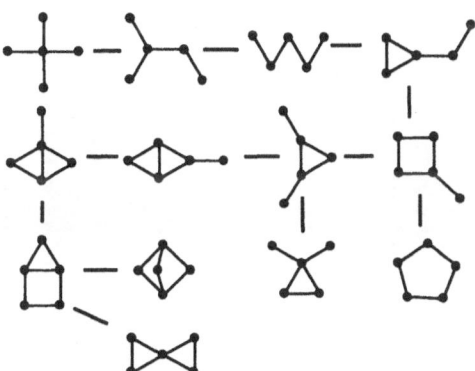

Fig. 19. The spanning tree of twelve hydrocarbons obtained by the similarity analysis of the fragment spectra weighted by the sum of degree of the subgraph nodes. The similarity value in Eq (1) was used for the analysis

Nonetheless, to evaluate structural similarity of compounds involving unsaturated bonds and heteroatoms it is necessary to express these differences properly at the stage of characterizing the structural fragment.

Figure 20 shows the fragment spectra for five psychotropic agents of which chemical structures are similar, when the fragments are characterized by the fragment weight. The resolution of the spectra is substantially increased by this fragment characterization. A result of evaluation of structural similarity of the molecules using the spectra in Fig. 20 is illustrated in Fig. 21.

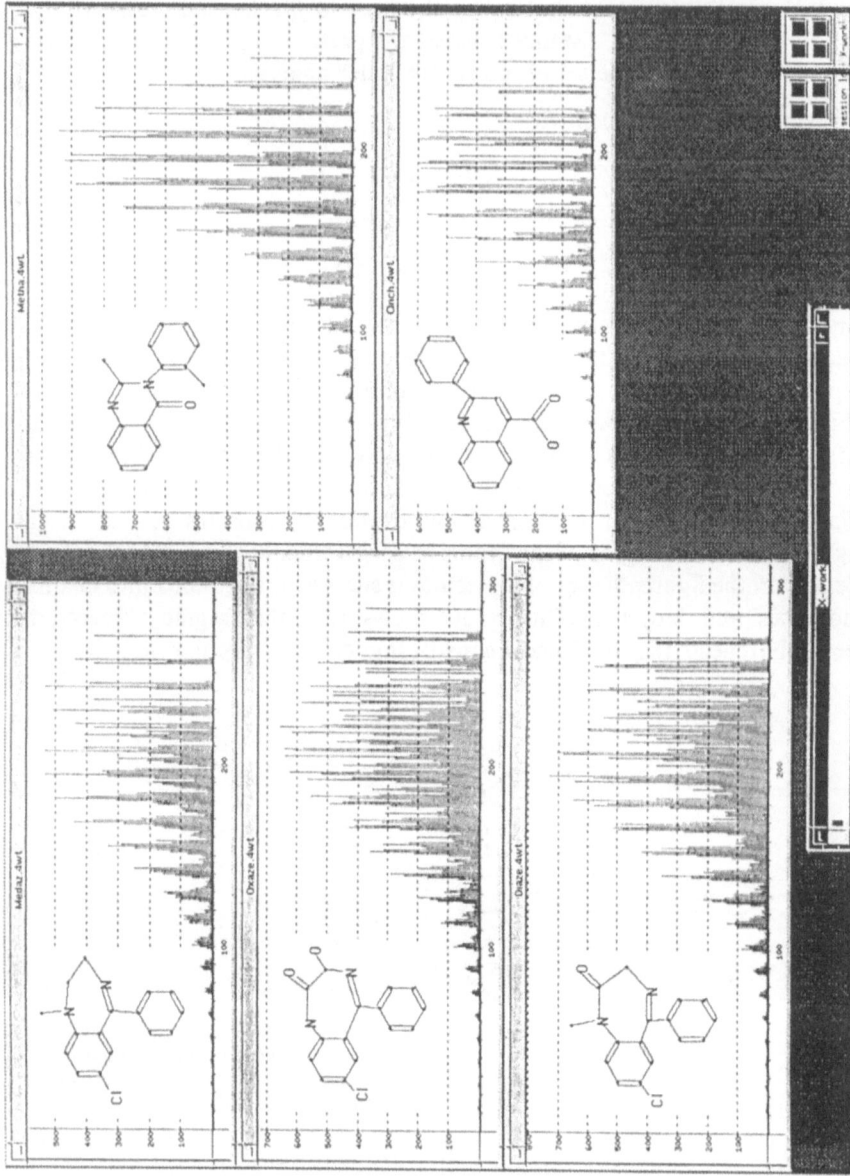

Fig. 20. Fragment spectra of five psycotropic agents that were weighted by substructural fragment mass

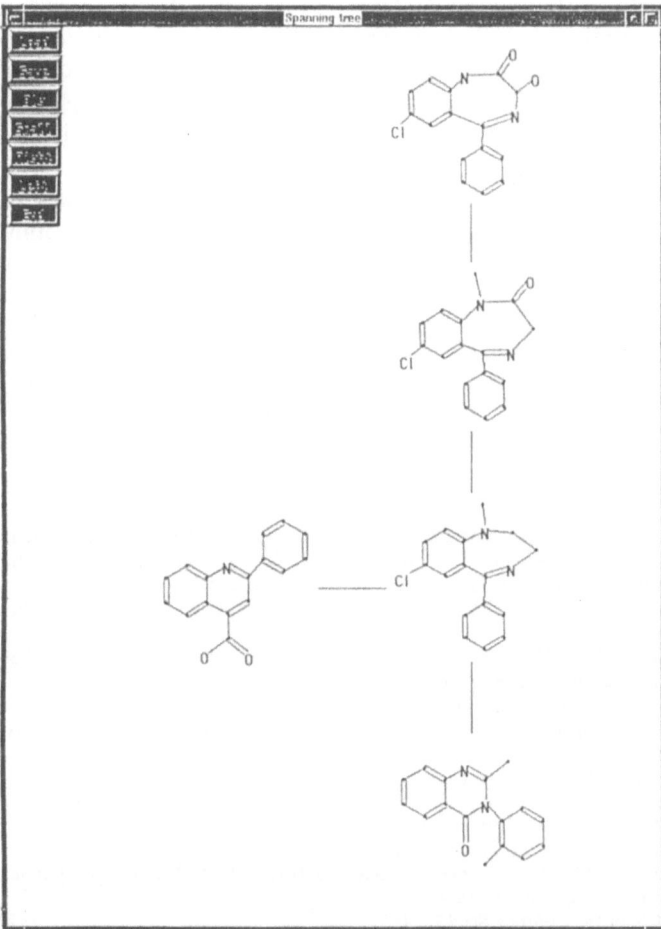

Fig. 21. The spanning tree of five psycotropic agents obtained by the similarity analysis of the fragment spectra in Fig. 20. The similarity value in Eq. (1) was used for the analysis

6 Conclusions

The identification and evaluation of the similarity of chemical structures in an objective and quantitative way is one of goals of chemical structure handling by computer. Conventionally, the problem of structural similarity of chemical compounds has been solved by chemists in an essentially intuitive way based on visual perception. The methods discussed herein are different from conventional substructure search technique. As already stated, the problem of computer handling of chemical structures is not only of the utmost importance in attempts to use computers in chemistry and related areas but it is also specific to each of those areas. It is believed that these approaches to evaluate structural similarities of chemical compounds are indispensable regarding higher applications of computers in chemistry and at the same time they constitute new techniques for chemical information processing which are necessarily required in this field.

Acknowledgement. The author thank the Science and Technology Agency of Japan for the valuable support by Special Coordination Funds for Promoting Science and Technology.

7 References

1. Wipke WT, Heller S, Feldman, Hyde E (eds) (1974) Computer representation and manipulation of chemical information. Wiley, New York
2. Ash JE, Hyde E (eds) (1975) Chemical information systems. Ellis Horwood, Chichester, England
3. Ash JE, Chubb PA, Ward SE, Welford SM, Willett P (1985) Communication, storage and retrieval of chemical information. Ellis Horwood, Chichester, England
4. Willett P (1987) Similarity and clustering in chemical information systems. Research Studies Press, Letchworth, England
5. Golender VE, Rozenblit AB (1983) Logical and combinatorial algorithms for drug design. Research Studies Press, Letchworth, England
6. Takahashi Y (1987) J Synth Org Chem Japan 45: 1087
7. Johnson MA, Maggiora GM (eds) (1990) Concepts and applications of molecular similarity. Wiley-Interscience, New York
8. Horvath AL (1992) Molecular design. Elsevier, Amsterdam
9. Adamson GW, Lynch MF, Town WG (1971) J Chem Soc C: 3702; Adamson GW, Lambourne DR, Lynch MF (1972) J Chem Soc Perkin Trans. 1: 2428
10. Adamson GW, Bush JA (1974) Nature 248: 406; (1975) J Chem Inf Comput Sci 15: 215
11. Adamson GW, Bawden D (1976) J Chem Inf Comput Sci 16: 161; (1977) J Chem Inf Comput Sci. 17: 164; (1980) J Chem Inf Comput Sci 20: 97
12. Bawden D (1983) J Chem Inf Comput Sci 23: 14
13. Willett P, Winterman V, Bawden P (1986) J Chem Inf Comput Sci 26: 109
14. Adamson GW, Bush JA (1975) J Chem Inf Comput Sci 15: 55
15. Carhart RE, Smith DH, Venkataraghavan (1985) J Chem Inf Comput Sci 25: 64
16. Klopman G (1984) J Am Chem Soc 106: 7315
17. Bawden D (1990) In: Johnson MA, Maggiora GM (eds) Concepts and applications of molecular similarity. Wiley-Interscience, New York, p 65
18. Kaltenbach TF, Small GW (1990) J Chem Inf Comput Sci 30: 73
19. Takahashi Y, Satoh Y, Suzuki H, Abe H, Sasaki S (1986) Anal Sci 2: 321
20. Judson (1992) J Chem Inf Comput Sci 32: 657
21. Hippe Z (1991) Artificial intelligence in chemistry, Elsevier, Amsterdam
22. Downs GM, Gillet VJ, Holliday JD, Lynch MF (1989) J Chem Inf Comput Sci 29: 172
23. Plotkin M (1971) J Chem Doc 11: 60
24. Downs GM, Gillet VJ, Holliday JD, Lynch MF (1989) J Chem Inf Comput Sci 29: 187; 29: 207; 29: 215
25. Takahashi Y (1994) J Chem Inf Comput Sci 34: 167
26. Armitage JE, Crowe JE, Evans PN, Lynch MF, McGuirk (1967) J Chem Doc 7: 209
27. Cone MM, Venkataraghavan R, McLafferty FW (1977) J Am Chem Soc 99: 7668
28. Varkony TH, Shiloach Y, Smith DH (1979) J Chem Inf Comput Sci 19: 104
29. Hyde E, Mattews FW, Thomson LH, Wiswesser WJ (1967) J Chem Doc 7: 200
30. Smith EG, Baker PA (1975) The Wisswesser line-formula chemical notation (WLN), Chemical Information Management Inc., Cherry Hill, New Jersey; Vollmer J (1983) J Chem Ed 60: 192

31. Takahashi Y, Satoh Y, Suzuki H, Sasaki S (1987) Anal Sci 3: 23
32. Bayada DM, Simpson RW, Johnson AP, Laurenco C. (1992) J Chem Inf Comput Sci 32: 680
33. Cohen NC (1985) Adv Drug Research 14: 41
34. Dean PM (ed) (1987) Molecular foundations of drug-receptor interaction. Cambridge Univ. Press, Cambridge
35. Richards WG (ed) (1989) Computer Aided Molecular Design, VCH publishers, New york
36. Carlos DA, Takahashi Y, Sasaki S (1993) J Mol Graphics 11: 23; (1993) J Chem Inf Comput Sci 33: 769
37. Gund P (1977) In: Hahn FE (ed), Progress in molecular and subcellular biology, vol 5. Springer, Berlin Heidelberg New York, p 117
38. Esaki T (1982) Chem Pharm Bull 30: 3657
39. Brint AT, Willett P (1987) J Mol Graphics 5: 49
40. Martin Y, Danaher EB, May CS, Weininger D (1988) J Comput Aided Mol Des 2: 15
41. Van Drie JH, Weininger D, Martin YC (1989) J Comput -Aided Mol Des 3: 225
42. Clark DE, Willett P (1992) J Mol Graphics 10: 194
43. Fisanick R, Cross KP, Forman JC, Rusinko A (1993) J Chem Inf Comput Sci 33: 548
44. Marshall GR, Barry CD, Bosshard HE, Dammkoehler RA, Dunn DA (1979) In: Olson EC, Christoffersen RE (eds) Computer-assisted drug design. ACS, Washington, DC, p 205
45. Hopfinger AJ (1980) J Am Chem Soc 120: 7196
46. Hopfinger AJ (1983) J Med Chem 26: 990
47. Jakes SE, Willett P (1986) J Mol Graphics 4: 12; Jakes SE, Watts NJ, Willett P, Bawden D, Fisher JD (1987) J Mol Graphics 5: 41
48. Marsili M, Floersheim P, Dreiding AS (1983) Comput & Chem 7: 175
49. Crippen GM (1982) Molec Phramcol 22: 11
50. Crandell CW, Smith DH (1983) J Chem Inf Comput Sci 23: 186
51. Danzinger DJ, Dean PM (1985) J Theor Biol 116: 215
52. Sheriden RP, Nilakantan R, Dixon JS, Venkataraghavan R (1986) J Med Chem 29: 899
53. Brint AT, Willett P (1987) J Chem Inf Comput Sci 27: 152
54. Takahashi Y, Maeda S, Sasaki S (1987) Anal Chim Acta 200: 363
55. Crippen GM, Havel TH (1979) J Med Chem 22: 988
56. Crippen GM, Havel TF (1988) Distance geometry and molecular confirmation. Research Studies Press, Somerset, England
57. Kuhl FS, Crippen GM, Friesen DK (1984) J Comp Chem 5: 24
58. Harary FS (1969) Graph Theory. Addison-Wesley, Reading
59. Michel AG, Evrard G, Schiltz M, Durant F, Koch MHJ (1976) Acta Chrystallogr. Sect B 32: 2507
60. Koch MHJ, Declercq JP, Germain G, Meerssche M (1977) Acta Crystallogr. Sect B 33: 2945
61. Morita K, Oka Y (1979) Kgaku: Zokan 79: 141
62. Takahashi Y, Sukekawa, Sasaki S (1992) J Chem Inf Comput Sci 32: 639
63. Cramer RD, Redl G, Berkoff CE (1974) J Med Chem 17: 535
64. Willett P, Winterman V (1986) Quant Struct-Act Relat 5: 18
65. Willett P (1987) Similarity and clustering technique in chemical information systems. Research Studies Press, Letchworth, England
66. Kier LB, Hall LH (1976) Molecular connectivity in chemistry and drug research. Academic Press, New York
67. Randic M (1990) In: Johnson MA, Maggiora GM (eds) Concepts and applications of molecular similarity. Wiley-Interscience, New York
68. Randic M (1991) J Math Chem 17: 155; (1992) J Chem Inf Comput Sci 32: 686
69. Weiner H (1947) J Am Chem Soc 69: 2636
70. Hosaya H (1971) Bull Chem Soc Japan 44: 2332
71. Randic M, Wilkins CL (1979) J Chem Inf Comput Sci 19: 23; 19: 31
72. Randic M, Wilkins CL (1979) Int J Quantum Chem Quantum Biol Symp 6: 55
73. Kier LB, Hall LH (1986) Molecular connectivity in structure–activity analysis. Wiley, New York
74. Horvath AL, Chapter 1 in ref. [8]
75. Takahashi Y, Ishiyama Y (1993) Abstracts of first girona seminar on molecular similarity. Girona, Spain

Author Index Volumes 151-174

Author Index Vols. 26-50 see Vol. 50
Author Index Vols. 51-100 see Vol. 100
Author Index Vols. 101-150 see Vol. 150

The volume numbers are printed in italics

Adam, W. and Hadjiarapoglou, L.: Dioxiranes: Oxidation Chemistry Made Easy. *164*, 45-62 (1993).

Albini, A., Fasani, E. and Mella M.: PET-Reactions of Aromatic Compounds. *168*, 143-173 (1993).

Allan, N.L. and Cooper, D.: Momentum-Space Electron Densities and Quantum Molecular Similarity. *173*, 85-111 (1995).

Allamandola, L.J.: Benzenoid Hydrocarbons in Space: The Evidence and Implications. *153*, 1-26 (1990).

Artymiuk, P. J., Poirette, A. R., Rice, D. W., and Willett, P.: The Use of Graph Theoretical Methods for the Comparison of the Structures of Biological Macromolecules. *174,* 73-104 (1995).

Astruc, D.: The Use of π-Organoiron Sandwiches in Aromatic Chemistry. *160*, 47-96 (1991).

Balzani, V., Barigelletti, F., De Cola, L.: Metal Complexes as Light Absorption and Light Emission Sensitizers. *158*, 31-71 (1990).

Baker, B.J. and Kerr, R.G.: Biosynthesis of Marine Sterols. *167*, 1-32 (1993).

Barigelletti, F., see Balzani, V.: *158*, 31-71 (1990).

Baumgarten, M., and Müllen, K.: Radical Ions: Where Organic Chemistry Meets Materials Sciences. *169*, 1-104 (1994).

Bersier, J., see Bersier, P.M.: *170*, 113-228 (1994).

Bersier, P. M., Carlsson, L., and Bersier, J.: Electrochemistry for a Better Environment. *170*, 113-228 (1994).

Besalú, E., Carbó, R., Mestres, J. and Solà, M.: Foundations and Recent Developments on Molecular Quantum Similarity. *173*, 31-62 (1995).

Bignozzi, C.A., see Scandola, F.: *158*, 73-149 (1990).

Billing, R., Rehorek, D., Hennig, H.: Photoinduced Electron Transfer in Ion Pairs. *158*, 151-199 (1990).

Bissell, R.A., de Silva, A.P., Gunaratne, H.Q.N., Lynch, P.L.M., Maguire, G.E.M., McCoy, C.P. and Sandanayake, K.R.A.S.: Fluorescent PET (Photoinduced Electron Transfer) Sensors. *168*, 223-264 (1993).

Blasse, B.: Vibrational Structure in the Luminescence Spectra of Ions in Solids. *171*, 1-26 (1994).

Bley, K., Gruber, B., Knauer, M., Stein, N. and Ugi, I.: New Elements in the Representation of the Logical Structure of Chemistry by Qualitative Mathematical Models and Corresponding Data Structures. *166*, 199-233 (1993).

Brunvoll, J., see Chen, R.S.: *153*, 227-254 (1990).

Brunvoll, J., Cyvin, B.N., and Cyvin, S.J.: Benzenoid Chemical Isomers and Their Enumeration. *162*, 181-221 (1992).

Brunvoll, J., see Cyvin, B.N.: *162*, 65-180 (1992).

Brunvoll, J., see Cyvin, S.J.: *166*, 65-119 (1993).

Bundle, D.R.: Synthesis of Oligosaccharides Related to Bacterial O-Antigens. *154*, 1-37 (1990).

Burrell, A.K., see Sessler, J.L.: *161*, 177-274 (1991).

Caffrey, M.: Structural, Mesomorphic and Time-Resolved Studies of Biological Liquid Crystals and Lipid Membranes Using Synchrotron X-Radiation. *151*, 75-109 (1989).

Canceill, J., see Collet, A.: *165*, 103-129 (1993).

Carbó, R., see Besalú, E.: *173*, 31-62 (1995).

Carlson, R., and Nordhal, A.: Exploring Organic Synthetic Experimental Procedures. *166*, 1-64 (1993).

Carlsson, L., see Bersier, P.M.: *170*, 113-228 (1994).

Ceulemans, A.: The Doublet States in Chromium (III) Complexes. A Shell-Theoretic View. *171*, 27-68 (1994).

Cimino, G. and Sodano, G.: Biosynthesis of Secondary Metabolites in Marine Molluscs. *167*, 77-116 (1993).

Chambron, J.-C., Dietrich-Buchecker, Ch., and Sauvage, J.-P.: From Classical Chirality to Topologically Chiral Catenands and Knots. *165*, 131-162 (1993).

Chang, C.W.J., and Scheuer, P.J.: Marine Isocyano Compounds. *167*, 33-76 (1993).

Chen, R.S., Cyvin, S.J., Cyvin, B.N., Brunvoll, J., and Klein, D.J.: Methods of Enumerating Kekulé Structures. Exemplified by Applified by Applications of Rectangle-Shaped Benzenoids. *153*, 227-254 (1990).

Chen, R.S., see Zhang, F.J.: *153*, 181-194 (1990).

Chiorboli, C., see Scandola, F.: *158*, 73-149 (1990).

Ciolowski, J.: Scaling Properties of Topological Invariants. *153*, 85-100 (1990).

Collet, A., Dutasta, J.-P., Lozach, B., and Canceill, J.: Cyclotriveratrylenes and Cryptophanes: Their Synthesis and Applications to Host-Guest Chemistry and to the Design of New Materials. *165*, 103-129 (1993).

Colombo, M. G., Hauser, A., and Güdel, H. U.: Competition Between Ligand Centered and Charge Transfer Lowest Excited States in bis Cyclometalated Rh^{3+} and Ir^{3+} Complexes. *171*, 143-172 (1994).

Cooper, D.L., Gerratt, J., and Raimondi, M.: The Spin-Coupled Valence Bond Description of Benzenoid Aromatic Molecules. *153*, 41-56 (1990).

Cooper, D.L., see Allan, N.L.: *173*, 85-111 (1995).

Cyvin, B.N., see Chen, R.S.: *153*, 227-254 (1990).

Cyvin, S.J., see Chen, R.S.: *153*, 227-254 (1990).

Cyvin, B.N., Brunvoll, J. and Cyvin, S.J.: Enumeration of Benzenoid Systems and Other Polyhexes. *162*, 65-180 (1992).

Cyvin, S.J., see Cyvin, B.N.: *162*, 65-180 (1992).

Cyvin, B.N., see Cyvin, S.J.: *166*, 65-119 (1993).

Cyvin, S.J., Cyvin, B.N., and Brunvoll, J.: Enumeration of Benzenoid Chemical Isomers with a Study of Constant-Isomer Series. *166*, 65-119 (1993).

Dartyge, E., see Fontaine, A.: *151*, 179-203 (1989).

De Cola, L., see Balzani, V.: *158*, 31-71 (1990).

Dear, K.: Cleaning-up Oxidations with Hydrogen Peroxide. *164*, (1993).

de Silva, A.P., see Bissell, R.A.: *168*, 223-264 (1993).

Descotes, G.: Synthetic Saccharide Photochemistry. *154*, 39-76 (1990).

Dias, J.R.: A Periodic Table for Benzenoid Hydrocarbons. *153*, 123-144 (1990).

Dietrich-Buchecker, Ch., see Chambron, J.-C.: *165*, 131-162 (1993).

Dohm, J., Vögtle, F.: Synthesis of (Strained) Macrocycles by Sulfone Pyrolysis. *161*, 69-106 (1991).

Dutasta, J.-P., see Collet, A.: *165*, 103-129 (1993).

Eaton, D.F.: Electron Transfer Processes in Imaging. *156*, 199-226 (1990).

El-Basil, S.: Caterpillar (Gutman) Trees in Chemical Graph Theory. *153*, 273-290 (1990).

Fasani, A., see Albini, A.: *168*, 143-173 (1993).

Fontaine, A., Dartyge, E., Itie, J.P., Juchs, A., Polian, A., Tolentino, H., and Tourillon, G.: Time-Resolved X-Ray Absorption Spectroscopy Using an Energy Dispensive Optics: Strengths and Limitations. *151*, 179-203 (1989).

Foote, C.S.: Photophysical and Photochemical Properties of Fullerenes. *169*, 347-364 (1994).

Fossey, J., Sorba, J., and Lefort, D.: Peracide and Free Radicals: A Theoretical and Experimental Approach. *164*, 99-113 (1993).

Fox, M.A.: Photoinduced Electron Transfer in Arranged Media. *159*, 67-102 (1991).

Freeman, P.K., and Hatlevig, S.A.: The Photochemistry of Polyhalocompounds, Dehalogenation by Photoinduced Electron Transfer, New Methods of Toxic Waste Disposal. *168*, 47-91 (1993).

Fuchigami, T.: Electrochemical Reactions of Fluoro Organic Compounds. *170*, 1-38 (1994).

Fuller, W., see Grenall, R.: *151*, 31-59 (1989).

Gehrke, R.: Research on Synthetic Polymers by Means of Experimental Techniques Employing Synchrotron Radiation. *151*, 111-159 (1989).

Gerratt, J., see Cooper, D.L.: *153*, 41-56 (1990).

Gerwick, W.H., Nagle, D.G., and Proteau, P.J.: Oxylipins from Marine Invertebrates. *167*, 117-180 (1993).

Gigg, J., and Gigg, R.: Synthesis of Glycolipids. *154*, 77-139 (1990).

Gislason, E.A., see Guyon, P.-M.: *151*, 161-178 (1989).

Greenall, R., Fuller, W.: High Angle Fibre Diffraction Studies on Conformational Transitions DNA Using Synchrotron Radiation. *151*, 31-59 (1989).

Gruber, B., see Bley, K.: *166*, 199-233 (1993).

Güdel, H. U., see Colombo, M. G.: *171*, 143-172 (1994).

Gunaratne, H.Q.N., see Bissell, R.A.: *168*, 223-264 (1993).

Guo, X.F., see Zhang, F.J.: *153*, 181-194 (1990).

Gust, D., and Moore, T.A.: Photosynthetic Model Systems. *159*, 103-152 (1991).

Gutman, I.: Topological Properties of Benzenoid Systems. *162*, 1-28 (1992).

Gutman, I.: Total π-Electron Energy of Benzenoid Hydrocarbons. *162*, 29-64 (1992).

Guyon, P.-M., Gislason, E.A.: Use of Synchrotron Radiation to Study-Selected Ion-Molecule Reactions. *151*, 161-178 (1989).

Hadjiarapoglou, L., see Adam, W.: *164*, 45-62 (1993).

Hart, H., see Vinod, T. K.: *172*, 119-178 (1994).

Harbottle, G.: Neutron Acitvation Analysis in Archaecological Chemistry. *157*, 57-92 (1990).

Hatlevig, S.A., see Freeman, P.K.: *168*, 47-91 (1993).

Hauser, A., see Colombo, M. G.: *171*, 143-172 (1994).

He, W.C., and He, W.J.: Peak-Valley Path Method on Benzenoid and Coronoid Systems. *153*, 195-210 (1990).

He, W.J., see He, W.C.: *153*, 195-210 (1990).

Heaney, H.: Novel Organic Peroxygen Reagents for Use in Organic Synthesis. *164*, 1-19 (1993).

Heinze, J.: Electronically Conducting Polymers. *152*, 1-19 (1989).

Helliwell, J., see Moffat, J.K.: *151*, 61-74 (1989).

Hennig, H., see Billing, R.: *158*, 151-199 (1990).

Hesse, M., see Meng, Q.: *161*, 107-176 (1991).

Hiberty, P.C.: The Distortive Tendencies of Delocalized π Electronic Systems. Benzene, Cyclobutadiene and Related Heteroannulenes. *153*, 27-40 (1990).

Hladka, E., Koca, J., Kratochvil, M., Kvasnicka, V., Matyska, L., Pospichal, J., and Potucek, V.: The Synthon Model and the Program PEGAS for Computer Assisted Organic Synthesis. *166*, 121-197 (1993).

Ho, T.L.: Trough-Bond Modulation of Reaction Centers by Remote Substituents. *155*, 81-158 (1990).

Höft, E.: Enantioselective Epoxidation with Peroxidic Oxygen. *164*, 63-77 (1993).

Hoggard, P. E.: Sharp-Line Electronic Spectra and Metal-Ligand Geometry. *171*, 113-142 (1994).

Holmes, K.C.: Synchrotron Radiation as a source for X-Ray Diffraction-The Beginning.*151*, 1-7 (1989).

Hopf, H., see Kostikov, R.R.: *155,* 41-80 (1990).

Indelli, M.T., see Scandola, F.: *158*, 73-149 (1990).

Inokuma, S., Sakai, S., and Nishimura, J.: Synthesis and Inophoric Properties of Crownophanes. *172*, 87-118 (1994).

Itie, J.P., see Fontaine, A.: *151*, 179-203 (1989).

Ito, Y.: Chemical Reactions Induced and Probed by Positive Muons. *157*, 93-128 (1990).

John, P., and Sachs, H.: Calculating the Numbers of Perfect Matchings and of Spanning Tress, Pauling's Bond Orders, the Characteristic Polynomial, and the Eigenvectors of a Benzenoid System. *153*, 145-180 (1990).

Jucha, A., see Fontaine, A.: *151*, 179-203 (1989).

Kaim, W.: Thermal and Light Induced Electron Transfer Reactions of Main Group Metal Hydrides and Organometallics. *169*, 231-252 (1994).

Kavarnos, G.J.: Fundamental Concepts of Photoinduced Electron Transfer. *156*, 21-58 (1990).

Kerr, R.G., see Baker, B.J.: *167*, 1-32 (1993).

Khairutdinov, R.F., see Zamaraev, K.I.: *163*, 1-94 (1992).

Kim, J.I., Stumpe, R., and Klenze, R.: Laser-induced Photoacoustic Spectroscopy for the Speciation of Transuranic Elements in Natural Aquatic Systems. *157*, 129-180 (1990).

Klaffke, W., see Thiem, J.: *154*, 285-332 (1990).

Klein, D.J.: Semiempirical Valence Bond Views for Benzenoid Hydrocarbons. *153*, 57-84 (1990).

Klein, D.J., see Chen, R.S.: *153*, 227-254 (1990).

Klenze, R., see Kim, J.I.: *157*, 129-180 (1990).

Knauer, M., see Bley, K.: *166*, 199-233 (1993).

Knops, P., Sendhoff, N., Mekelburger, H.-B., Vögtle, F.: High Dilution Reactions - New Synthetic Applications. *161*, 1-36 (1991).

Koca, J., see Hladka, E.: *166*, 121-197 (1993).

Koepp, E., see Ostrowicky, A.: *161*, 37-68 (1991).

Kohnke, F.H., Mathias, J.P., and Stoddart, J.F.: Substrate-Directed Synthesis: The Rapid Assembly of Novel Macropolycyclic Structures *via* Stereoregular Diels-Alder Oligomerizations. *165*, 1-69 (1993).

Kostikov, R.R., Molchanov, A.P., and Hopf, H.: Gem-Dihalocyclopropanos in Organic Synthesis. *155*, 41-80 (1990).

Kratochvil, M., see Hladka, E.: *166*, 121-197 (1993).

Kumar, A., see Mishra, P. C.: *174*, 27-44 (1995).

Krogh, E., and Wan, P.: Photoinduced Electron Transfer of Carbanions and Carbacations. *156*, 93-116 (1990).

Kunkeley, H., see Vogler, A.: *158*, 1-30 (1990).

Kuwajima, I., and Nakamura, E.: Metal Homoenolates from Siloxycyclopropanes. *155*, 1-39 (1990).

Kvasnicka, V., see Hladka, E.: *166*, 121-197 (1993).

Lange, F., see Mandelkow, E.: *151*, 9-29 (1989).

Lefort, D., see Fossey, J.: *164*, 99-113 (1993).

Lopez, L.: Photoinduced Electron Transfer Oxygenations. *156*, 117-166 (1990).

Lozach, B., see Collet, A.: *165*, 103-129 (1993).

Lymar, S.V., Parmon, V.N., and Zamarev, K.I.: Photoinduced Electron Transfer Across Membranes. *159*, 1-66 (1991).

Lynch, P.L.M., see Bissell, R.A.: *168*, 223-264 (1993).

Maguire, G.E.M., see Bissell, R.A.: *168*, 223-264 (1993).

Mandelkow, E., Lange, G., Mandelkow, E.-M.: Applications of Synchrotron Radiation to the Study of Biopolymers in Solution: Time-Resolved X-Ray Scattering of Microtubule Self-Assembly and Oscillations. *151*, 9-29 (1989).

Mandelkow, E.-M., see Mandelkow, E.: *151*, 9-29 (1989).

Maslak, P.: Fragmentations by Photoinduced Electron Transfer. Fundamentals and Practical Aspects. *168*, 1-46 (1993).

Mathias, J.P., see Kohnke, F.H.: *165*, 1-69 (1993).

Mattay, J., and Vondenhof, M.: Contact and Solvent-Separated Radical Ion Pairs in Organic Photochemistry. *159*, 219-255 (1991).

Matyska, L., see Hladka, E.: *166*, 121-197 (1993).

McCoy, C.P., see Bissell, R.A.: *168*, 223-264 (1993).

Mekelburger, H.-B., see Knops, P.: *161*, 1-36 (1991).

Mekelburger, H.-B., see Schröder, A.: *172*, 179-201 (1994).

Mella, M., see Albini, A.: *168*, 143-173 (1993).

Memming, R.: Photoinduced Charge Transfer Processes at Semiconductor Electrodes and Particles. *169*, 105-182 (1994).

Meng, Q., Hesse, M.: Ring Closure Methods in the Synthesis of Macrocyclic Natural Products. *161*, 107-176 (1991).

Merz, A.: Chemically Modified Electrodes. *152*, 49-90 (1989).

Meyer, B.: Conformational Aspects of Oligosaccharides. *154*, 141-208 (1990).

Mishra, P. C., and Kumar A.: Mapping of Molecular Electric Potentials and Fields. *174*, 27-44 (1995).

Mestres, J., see Besalú, E.: *173*, 31-62 (1995).

Mezey, P.G.: Density Domain Bonding Topology and Molecular Similarity Measures. *173*, 63-83 (1995).

Misumi, S.: Recognitory Coloration of Cations with Chromoacerands. *165*, 163-192 (1993).

Mizuno, K., and Otsuji, Y.: Addition and Cycloaddition Reactions via Photoinduced Electron Transfer. *169*, 301-346 (1994).

Moffat, J.K., Helliwell, J.: The Laue Method and its Use in Time-Resolved Crystallography. *151*, 61-74 (1989).

Molchanov, A.P., see Kostikov, R.R.: *155*, 41-80 (1990).

Moore, T.A., see Gust, D.: *159*, 103-152 (1991).

Müllen, K., see Baumgarten, M.: *169*, 1-104 (1994).

Nagle, D.G., see Gerwick, W.H.: *167*, 117-180 (1993).

Nakamura, E., see Kuwajima, I.: *155*, 1-39 (1990).

Nishimura, J., see Inokuma, S.: *172*, 87-118 (1994).

Nordahl, A., see Carlson, R.: *166*, 1-64 (1993).

Okuda, J.: Transition Metal Complexes of Sterically Demanding Cyclopentadienyl Ligands. *160*, 97-146 (1991).

Ostrowicky, A., Koepp, E., Vögtle, F.: The "Vesium Effect": Synthesis of Medio- and Macrocyclic Compounds. *161*, 37-68 (1991).

Otsuji, Y., see Mizuno, K.: *169*, 301-346 (1994).

Pálinkó, I., see Tasi, G.: *174*, 45-72 (1995).

Pandey, G.: Photoinduced Electron Transfer (PET) in Organic Synthesis. *168*, 175-221 (1993).

Parmon, V.N., see Lymar, S.V.: *159*, 1-66 (1991).

Poirette, A. R., see Artymiuk, P. J.: *174*, 73-104 (1995).

Polian, A., see Fontaine, A.: *151*, 179-203 (1989).

Ponec, R.: Similarity Models in the Theory of Pericyclic Macromolecules. *174*, 1-26 (1995).

Pospichal, J., see Hladka, E.: *166*, 121-197 (1993).

Potucek, V., see Hladka, E.: *166*, 121-197 (1993).

Proteau, P.J., see Gerwick, W.H.: *167*, 117-180 (1993).

Raimondi, M., see Copper, D.L.: *153*, 41-56 (1990).

Reber, C., see Wexler, D.: *171*, 173-204 (1994).

Rettig, W.: Photoinduced Charge Separation via Twisted Intramolecular Charge Transfer States. *169*, 253-300 (1994).

Rice, D. W., see Artymiuk, P. J.: *174*, 73-104 (1995).

Riekel, C.: Experimental Possibilities in Small Angle Scattering at the European Synchrotron Radiation Facility. *151*, 205-229 (1989).

Roth, H.D.: A Brief History of Photoinduced Electron Transfer and Related Reactions. *156*, 1-20 (1990).

Roth, H.D.: Structure and Reactivity of Organic Radical Cations. *163*, 131-245 (1992).

Rouvray, D.H.: Similarity in Chemistry: Past, Present and Future. *173*, 1-30 (1995).

Rüsch, M., see Warwel, S.: *164*, 79-98 (1993).

Sachs, H., see John, P.: *153*, 145-180 (1990).

Saeva, F.D.: Photoinduced Electron Transfer (PET) Bond Cleavage Reactions. *156*, 59-92 (1990).

Sakai, S., see Inokuma, S.: *172*, 87-118 (1994).

Sandanayake, K.R.A.S., see Bissel, R.A.: *168*, 223-264 (1993).

Sauvage, J.-P., see Chambron, J.-C.: *165*, 131-162 (1993).

Schäfer, H.-J.: Recent Contributions of Kolbe Electrolysis to Organic Synthesis. *152*, 91-151 (1989).

Scheuer, P.J., see Chang, C.W.J.: *167*, 33-76 (1993).

Schmidtke, H.-H.: Vibrational Progressions in Electronic Spectra of Complex Compounds Indicating Stron Vibronic Coupling. *171*, 69-112 (1994).

Schmittel, M.: Umpolung of Ketones via Enol Radical Cations. *169*, 183-230 (1994).

Schröder, A., Mekelburger, H.-B., and Vögtle, F.: Belt-, Ball-, and Tube-shaped Molecules. *172*, 179-201 (1994).

Schulz, J., Vögtle, F.: Transition Metal Complexes of (Strained) Cyclophanes. *172*, 41-86 (1994).

Sendhoff, N., see Knops, P.: *161*, 1-36 (1991).

Sessler, J.L., Burrell, A.K.: Expanded Porphyrins. *161*, 177-274 (1991).

Sheldon, R.: Homogeneous and Heterogeneous Catalytic Oxidations with Peroxide Reagents. *164*, 21-43 (1993).

Sheng, R.: Rapid Ways of Recognize Kekuléan Benzenoid Systems. *153*, 211-226 (1990).

Sodano, G., see Cimino, G.: *167*, 77-116 (1993).

Sojka, M., see Warwel, S.: *164*, 79-98 (1993).

Solà, M., see Besalú, E.: *173*, 31-62 (1995).

Sorba, J., see Fossey, J.: *164*, 99-113 (1993).

Stanek, Jr., J.: Preparation of Selectively Alkylated Saccharides as Synthetic Intermediates. *154*, 209-256 (1990).

Steckhan, E.: Electroenzymatic Synthesis. *170*, 83-112 (1994).

Stein, N., see Bley, K.: *166*, 199-233 (1993).

Stoddart, J.F., see Kohnke, F.H.: *165*, 1-69 (1993).

Soumillion, J.-P.: Photoinduced Electron Transfer Employing Organic Anions. *168*, 93-141 (1993).

Stumpe, R., see Kim, J.I.: *157*, 129-180 (1990).

Suami, T.: Chemistry of Pseudo-sugars. *154*, 257-283 (1990).

Suppan, P.: The Marcus Inverted Region. *163*, 95-130 (1992).

Suzuki, N.: Radiometric Determination of Trace Elements. *157*, 35-56 (1990).

Takahashi, Y.: Identification of Structural Similarity of Organic Molecules. *174*, 105-134 (1995).

Tasi, G., and Pálinkó, I.: Using Molecular Electrostatic Potential Maps for Similarity Studies. *174*, 45-72 (1995).

Thiem, J., and Klaffke, W.: Synthesis of Deoxy Oligosaccharides. *154*, 285-332 (1990).

Timpe, H.-J.: Photoinduced Electron Transfer Polymerization. *156*, 167-198 (1990).

Tobe, Y.: Strained [n]Cyclophanes. *172*, 1-40 (1994.

Tolentino, H., see Fontaine, A.: *151*, 179-203 (1989).

Tomalia, D.A.: Genealogically Directed Synthesis: Starbust/Cascade Dendrimers and Hyperbranched Structures. *165*, (1993).

Tourillon, G., see Fontaine, A.: *151*, 179-203 (1989).

Ugi, I., see Bley, K.: *166*, 199-233 (1993).

Vinod, T. K., Hart, H.: Cuppedo- and Cappedophanes. *172*, 119-178 (1994).

Vögtle, F., see Dohm, J.: *161*, 69-106 (1991).

Vögtle, F., see Knops, P.: *161*, 1-36 (1991).

Vögtle, F., see Ostrowicky, A.: *161*, 37-68 (1991).

Vögtle, F., see Schulz, J.: *172*, 41-86 (1994).

Vögtle, F., see Schröder, A.: *172*, 179-201 (1994).

Vogler, A., Kunkeley, H.: Photochemistry of Transition Metal Complexes Induced by Outer-Sphere Charge Transfer Excitation. *158*, 1-30 (1990).

Vondenhof, M., see Mattay, J.: *159*, 219-255 (1991).

Wan, P., see Krogh, E.: *156*, 93-116 (1990).

Warwel, S., Sojka, M., and Rüsch, M.: Synthesis of Dicarboxylic Acids by Transition-Metal Catalyzed Oxidative Cleavage of Terminal-Unsaturated Fatty Acids. *164*, 79-98 (1993).

Wexler, D., Zink, J. I., and Reber, C.: Spectroscopic Manifestations of Potential Surface Coupling Along Normal Coordinates in Transition Metal Complexes. *171*, 173-204 (1994).

Willett, P., see Artymiuk, P. J.: *174*, 73-104 (1995).

Willner, I., and Willner, B.: Artificial Photosynthetic Model Systems Using Light-Induced Electron Transfer Reactions in Catalytic and Biocatalytic Assemblies. *159*, 153-218 (1991).

Yoshida, J.: Electrochemical Reactions of Organosilicon Compounds. *170*, 39-82 (1994).

Yoshihara, K.: Chemical Nuclear Probes Using Photon Intensity Ratios. *157*, 1-34 (1990).

Zamaraev, K.I., see Lymar, S.V.: *159*, 1-66 (1991).

Zamaraev, K.I., Kairutdinov, R.F.: Photoinduced Electron Tunneling Reactions in Chemistry and Biology. *163*, 1-94 (1992).

Zander, M.: Molecular Topology and Chemical Reactivity of Polynuclear Benzenoid Hydrocarbons. *153*, 101-122 (1990).

Zhang, F.J., Guo, X.F., and Chen, R.S.: The Existence of Kekulé Structures in a Benzenoid System. *153*, 181-194 (1990).

Zimmermann, S.C.: Rigid Molecular Tweezers as Hosts for the Complexation of Neutral Guests. *165*, 71-102 (1993).

Zink, J. I., see Wexler, D.: *171*, 173-204 (1994).

Zybill, Ch.: The Coordination Chemistry of Low Valent Silicon. *160*, 1-46 (1991).